# Digital Innovations in Architecture, Engineering and Construction

The Architecture, Engineering and Construction (AEC) industry is experiencing an unprecedented transformation from conventional labor-intensive activities to automation using innovative digital technologies and processes. This new paradigm also requires systemic changes focused on social, economic and sustainability aspects. Within the scope of Industry 4.0, digital technologies are a key factor in interconnecting information between the physical built environment and the digital virtual ecosystem. The most advanced virtual ecosystems allow to simulate the built to enable a real-time data-driven decision-making. This Book Series promotes and expedites the dissemination of recent research, advances, and applications in the field of digital innovations in the AEC industry. Topics of interest include but are not limited to:

- Industrialization: digital fabrication, modularization, cobotics, lean.
- Material innovations: bio-inspired, nano and recycled materials.
- Reality capture: computer vision, photogrammetry, laser scanning, drones.
- Extended reality: augmented, virtual and mixed reality.
- Sustainability and circular building economy.
- Interoperability: building/city information modeling.
- Interactive and adaptive architecture.
- Computational design: data-driven, generative and performance-based design.
- Simulation and analysis: digital twins, virtual cities.
- Data analytics: artificial intelligence, machine/deep learning.
- Health and safety: mobile and wearable devices, QR codes, RFID.
- Big data: GIS, IoT, sensors, cloud computing.
- Smart transactions, cybersecurity, gamification, blockchain.
- Quality and project management, business models, legal prospective.
- Risk and disaster management.

Lenka Kabošová · Dušan Katunský ·
Stanislav Kmet'

# Designing with the Wind

## Climate-Derived Architecture

Springer

Lenka Kabošová
CRIC–Center for Research and Innovation
in Construction
Technical University of Košice
Košice, Slovakia

Dušan Katunský
Faculty of Civil Engineering
Technical University of Košice
Košice, Slovakia

Stanislav Kmeť
Faculty of Civil Engineering
Technical University of Košice
Košice, Slovakia

ISSN 2731-7269 ISSN 2731-7277 (electronic)
Digital Innovations in Architecture, Engineering and Construction
ISBN 978-3-031-24440-7 ISBN 978-3-031-24441-4 (eBook)
https://doi.org/10.1007/978-3-031-24441-4



This Springer imprint is published by the registered company Springer Nature Switzerland AG
The registered company address is: Gewerbestrasse 11, 6330 Cham, Switzerland

*We wish to dedicate this book to all architecture, engineering, and science enthusiasts who incessantly search for inventive and sustainable ways of blending technology and nature. Our planet is unique and is our only home for the next x generations.*

*Special appreciation goes to our families, without which our determination and diligence for science work would not be possible.*

# Preface

Witnessing the climate changes, architects and civil engineers challenge the sustainability of the conventional building design approach. The pursuit for sustainable energy resources and the tendency for new materials and techniques to enhance the insulation properties of our houses is given great heed in the building industry, aiming for energy-plus buildings.

Throughout the past several decades, the architectural research has started centering on the design and creation of architecture, leveraging the benefits of the digital revolution yet utilizing the nature-driven design approach. Buildings that can sense their surroundings and respond through shape or material changes, adapting to changing ambient conditions, are another exciting research impulse.

With the enhancing hardware and software, design problems involving the interplay of technology, architecture, and nature, can be tackled. The factors such as solar radiation, fluctuating temperatures, wind pressure, movement of inhabitants, extreme climatic situations, such as an earthquake, and others, can be incorporated into the early conceptual stage and thus influence the design, form-finding, and optimization of architecture.

This book explores wind-adaptive architectural design blending the parametric design with digital simulations to suggest a novel approach for specific, even extreme, wind conditions. Ideally, this wind-related design method incorporated into the design routine will be the first step in creating architecture that can act in response to the nature around.

Košice, Slovakia

Lenka Kabošová
Dušan Katunský
Stanislav Kmet'

**Acknowledgements** This work was realized within the scope and with the financial support of the scientific projects VEGA 1/0302/16, VEGA 1/0674/18, VEGA 1/0374/19, VEGA 1/0129/20, and VEGA 1/0626/22 (Grant Agency of the Slovak Republic).

# About This Book

Digitalization and automatization in all fields of our lives are the key technological topics of the past decades. Although the threshold between the human and digital gradually blends, we are just starting to fully grasp the potential of the digital age for mitigating and (perhaps) reverting the more frequent and evident extreme weather events. The temperature changes, the rapid loss of ice, the occurrence of strong winds or wind gusts, floods, droughts, and other extremes impact our life as well as the environment [1]. Not surprisingly, the search for causes unveils that many environmental problems, including global warming, are related to the activities of humans [2, 3]. Although changes in the weather patterns are gradual, they have already impacted the architectural design. A question arises for planners, engineers, and architects: Could we design buildings in a way that reacts to this situation and maybe even reverts it? The environment, especially the weather, has an undeniable influence on architecture. And vice versa, leveraging architecture and urbanism, the local climate can be affected and consequentially improved. Weather-related design methods are of immense interest in the climate change era. By synthesizing the creativity and skills of the architect with digital designing, architecture can become an active player in its environment.

This book proposes an urban and architectural design that emerges from the specific wind microclimate of the design site and responds to the changes in the ambient wind conditions. The investigated topic is interdisciplinary, involving architecture, computer engineering, and civil and environmental engineering.

Amidst all the weather forces that affect architecture, the interaction of wind fluxes with building shape is the most perceptible. Considering the wind microclimate of the design site in the early conceptual stage contributes to creating environment-based buildings, as well as comfortable public spaces around them [4–6].

The adverse wind effects manifest as loads (pressure and suction on building facades) or outdoor pedestrian wind discomfort. The discomfort can be related to either high-speed wind fluxes in cold climates or absent wind in arid climates. Besides the wind analysis of high-rise buildings and skyscrapers, until recently, little attention has been given to the wind evaluation of building designs during their creation. Nonetheless, decades of the development of the Computational Fluid Dynamics

(CFD) simulations and the efforts for their integration into building engineering [7–9] have brought a prospect for examining the complex relations of the wind phenomenon in the context of architecture.

The focus is on the domain of wind-driven form-finding in the conceptual design stage. The book will look closely at (A) the interdisciplinary wind-driven design method for architects, engineers, and urbanists employing open-source software for CFD analysis and (B) the tensegrity-membrane adaptive building façades. The main research question throughout the book is:

**Can the wind-driven methodology enhance the wind comfort around buildings? Can it lead to the reduction of wind surface loads acting on buildings?**

# Contents

# About the Authors

**Arch. Lenka Kabošová, Ph.D.** is an architect and a researcher, focusing her research on exploring climate-oriented, wind-driven architectural design combining a parametric design approach with digital simulations to propose novel ways of designing for specific, even extreme wind conditions. Throughout the doctoral studies, fruitful research cooperations with various architectural Universities and Research Institutions in several European countries have led to publications in indexed journals and long-term collaboration. In 2019, her Ph.D. thesis was awarded the first prize among the selected doctoral students from all nine faculties of the Technical University in Košice.

**Prof. Dušan Katunský, CSc.** is a recognized researcher in building physics and dean of the Faculty of Civil Engineering at the Technical University in Košice. Prior to this, he was the head of the Institute of Architectural Engineering and the Institute of Building Technology. His research focuses on building physics, including heat and moisture analysis, aerodynamics, hydrodynamics, and acoustics. He has been leading and mentoring many successful Ph.D. candidates from home and abroad. He is the author of several well-cited and recognized scientific papers in indexed international journals. He is a reviewer for publishers such as Elsevier and Thomson Reuters. He acts as the editor-in-chief for the University's Selected Scientific Papers—Journal of Civil Engineering, which has been issued for 17 years. He is a qualified expert in construction and has completed several expert assessments for practice, courts, the police, and the prosecutor's office.

**Prof. Stanislav Kmet', DrSc., Dr.h.c., prof.h.c.,** is a renowned researcher in Structural Engineering and rector of the Technical University of Košice. His professional background is in the theory, design, and experimental research of large-span adaptive cable, membrane, and tensegrity systems with the application of artificial intelligence methods. His research interests are probabilistic reliability analysis, advanced time-dependent nonlinear mathematically-physical computational methods, elastic-plastic and rheological models, simulations, and behavioral modelings of structures and structural members subjected to static and dynamic load effects. He lectures as an invited speaker at foreign universities, conferences, symposiums, and seminars. He is an author and co-author of five books, over 110 scientific and professional journal papers, more than 240 articles in conference proceedings, 36 projects, and over 50 expertise reports for the industrial practice and others.

# Chapter 1
# Background

This chapter contains four sub-chapters, all of which create an essential interdisciplinary background for developing the wind-driven design method:

1. Extreme environment
2. The wind
3. Algorithms-Aided Design and simulations in architecture
4. Adaptive architecture.

## 1.1 Extreme Environment

Adjusting the building design to changing climate conditions is inevitable, still, the question of the adaptation level is open [1].

### *1.1.1 Definitions*

- **Extreme environment**. Natural and human-made (artificial) environments, in which extreme events prevail. They can either be events that exceed threshold values of certain important meteorological variables or events with a high impact on society or nature.
- **Global warming**. Due to the presence of greenhouse gases such as carbon dioxide, methane, and nitrous oxide in the atmosphere, most of which cumulate because of human activities, long-wave solar rays are not able to reflect from Earth's surface to the outer space and are retained in the atmosphere causing the global temperature to rise [2].
- **Climate change**. "A change in global or regional climate patterns, in particular, a change apparent from the mid to late twentieth century onwards and attributed

 
L. Kabošová et al., *Designing with the Wind*, Digital Innovations in Architecture, Engineering and Construction, https://doi.org/10.1007/978-3-031-24441-4_1

largely to the increased levels of atmospheric carbon dioxide generated by fossil fuels" [3].

- **Climate Reanalyzer (CCI-Reanalyzer)**. This website created by the Climate Change Institute at the University of Maine provides access to publicly available climatic datasets and models. The user can produce comparisons and predictions of various climate change indicators, such as temperature extremes, anomalies within wind speed or precipitation, or sea ice concentration.
- **The Intergovernmental Panel on Climate Change (IPCC)**. The United Nations non-governmental body for assessing the science related to climate change. It builds upon the scientific studies of hundreds of selected researchers.

### *1.1.2 Global Climate Change*

Referring to multiple research projects, NASA, collaborating with the Goddard Institute for Space Studies, concludes that the evidence of climate change is substantial and includes [4, 5]:

- The average global temperature has risen about 0.98 °C relative to the 1951–1980 average temperatures, which is caused by the increasing level of $CO_2$ and other human-made emissions.
- The oceans have warmed more than 0.33 °C since 1969.
- Sea levels have risen, reaching 20 cm during the last century. An additional 18–100 cm increase is expected by the year 2100 [2].
- Ice sheets and ice shelf mass have shrunk in Greenland and Antarctica. Glaciers are receding, particularly in the Alps, Himalayas, Andes, Rockies, Alaska, and Africa.
- The snow cover in the Northern hemisphere is decreasing during winter.
- Extreme events are occurring more, including more frequent extremely high-temperature events, less frequent extremely low-temperature events, and intense rainfalls producing massive floods.
- Oceans become more acidic while absorbing more $CO_2$ (an increase of about 30% since the beginning of the Industrial Revolution).

The IPCC findings, reflecting the global scientific consensus [2], reinforce NASA's conclusions in the assessment reports from 2014 and 2018 [6–8]:

- Warming of the climate system is unequivocal. Since the 1950s, many of the observed changes have been unprecedented over decades to millennia. The atmosphere and ocean have warmed, the amounts of snow and ice have diminished, and the sea level has risen.
- The IPCC has 95-percent certainty that humans are the primary cause of current global warming.

Figure 1.1 depicts the trend of increasing temperatures. The alarming fact is that the differences are apparent when comparing the last two decades. By the year 2050,

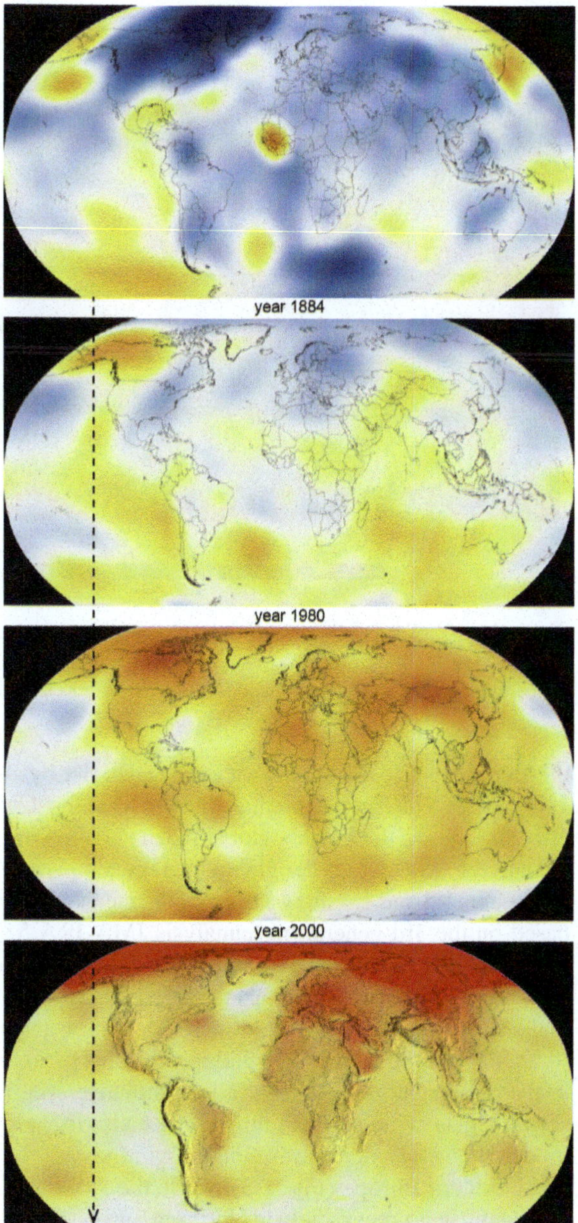


**Fig. 1.1** Global temperature in 1884, 1980, 2000, and 2021 relative to the average temperature in the years 1951–1980 (*Source* https://climate.nasa.gov/vital-signs/global-temperature/)

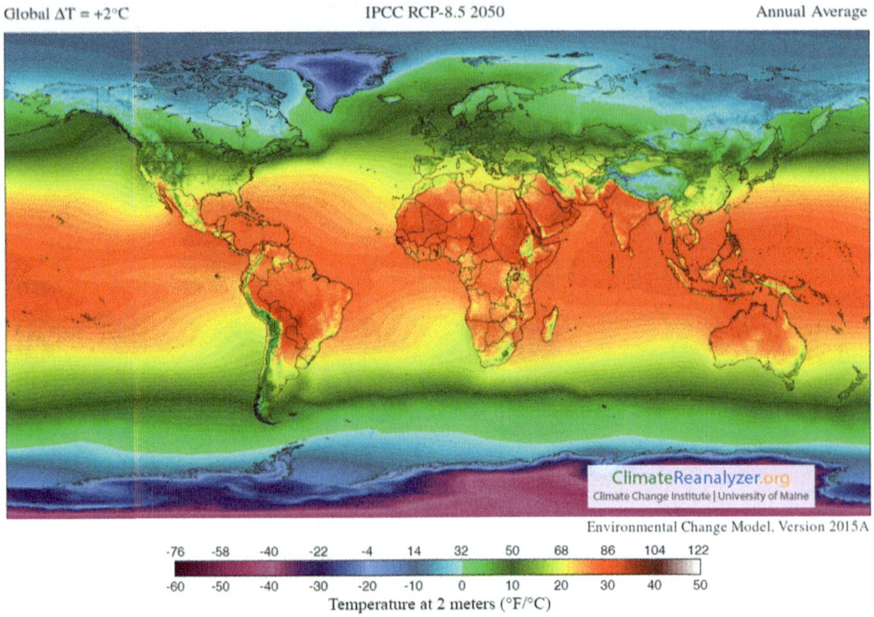


**Fig. 1.2** The 2 °C global mean surface temperature rise and the resulting annual average temperature at 2 m above the ground (*Source* Climate Reanalyzer)

the global average temperature is expected to rise by about 2 °C compared to the years 1979–2000. The Environmental Change Model (ECM) developed by the University of Maine shows the Earth's global mean surface temperature (GMST) at 2 m above the surface (Fig. 1.2) if the global average temperature will be 2 °C higher.

Figures 1.3 and 1.4 show a trend of the mean wind speeds at 10 m above the surface based on the 3rd generation reanalysis JMA JRA 55 model and the 1st generation NCEP/NCAR model [9]. Two time periods are juxtaposed: the years 1990–2020 with the years 1958–1989. The goal is to understand the development of the wind situation in the world over the past decades. The data show the trend of increasing wind speeds (orange to red) or decreasing wind speeds (blue to violet) in the last 30 years compared to the previous three decades. The cumulative comparison of the two time periods shows higher wind speeds in the last decades in various regions. The NCEP/NCAR model shows an increase in annual wind speeds of about 0.2–0.5 m/s in Central and Northern Europe (Fig. 1.3), while the JMA JRA 55 model shows the trend of increasing wind speeds only in Northern Europe (Fig. 1.4). Although CCI-Reanalyzer does not provide projections for the future development of the annual-mean near-surface wind speeds, we can conclude, based on the trends, that the tendency of stronger mean near-surface winds is apparent around poles and above oceans.

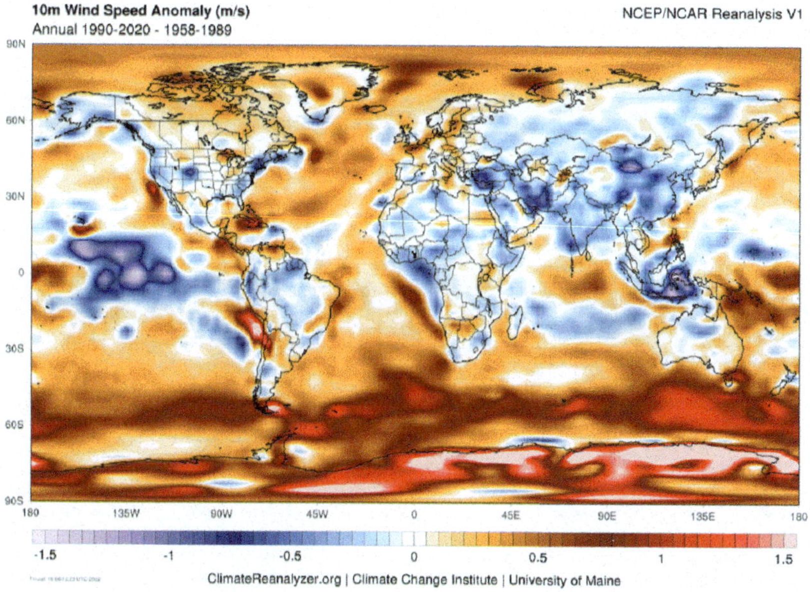


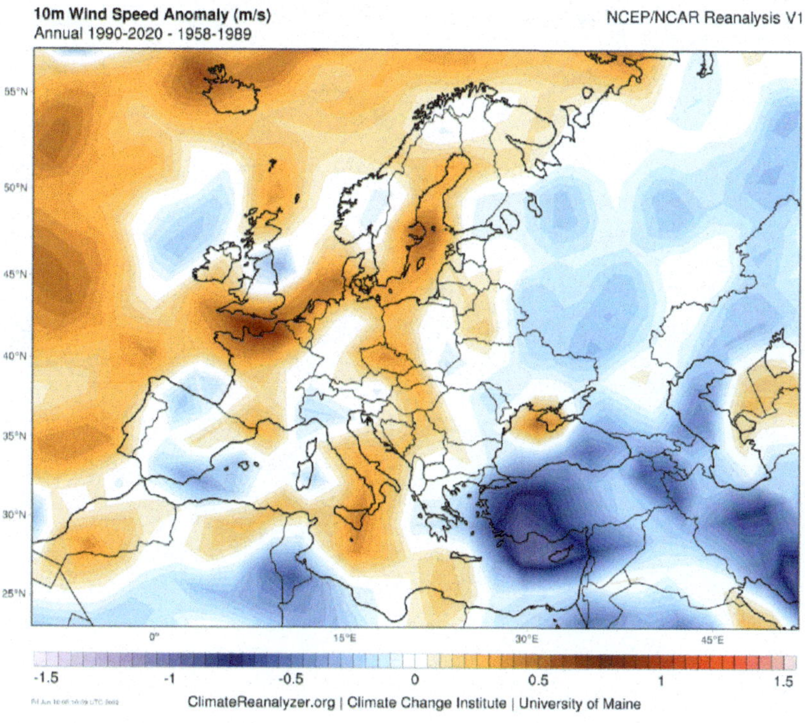


**Fig. 1.3** The juxtaposition of the mean wind speeds at 10 m above the ground in two time periods, 1990–2020 and 1958–1989. The NCEP/NCAR analysis model (*Source* Climate Reanalyzer)

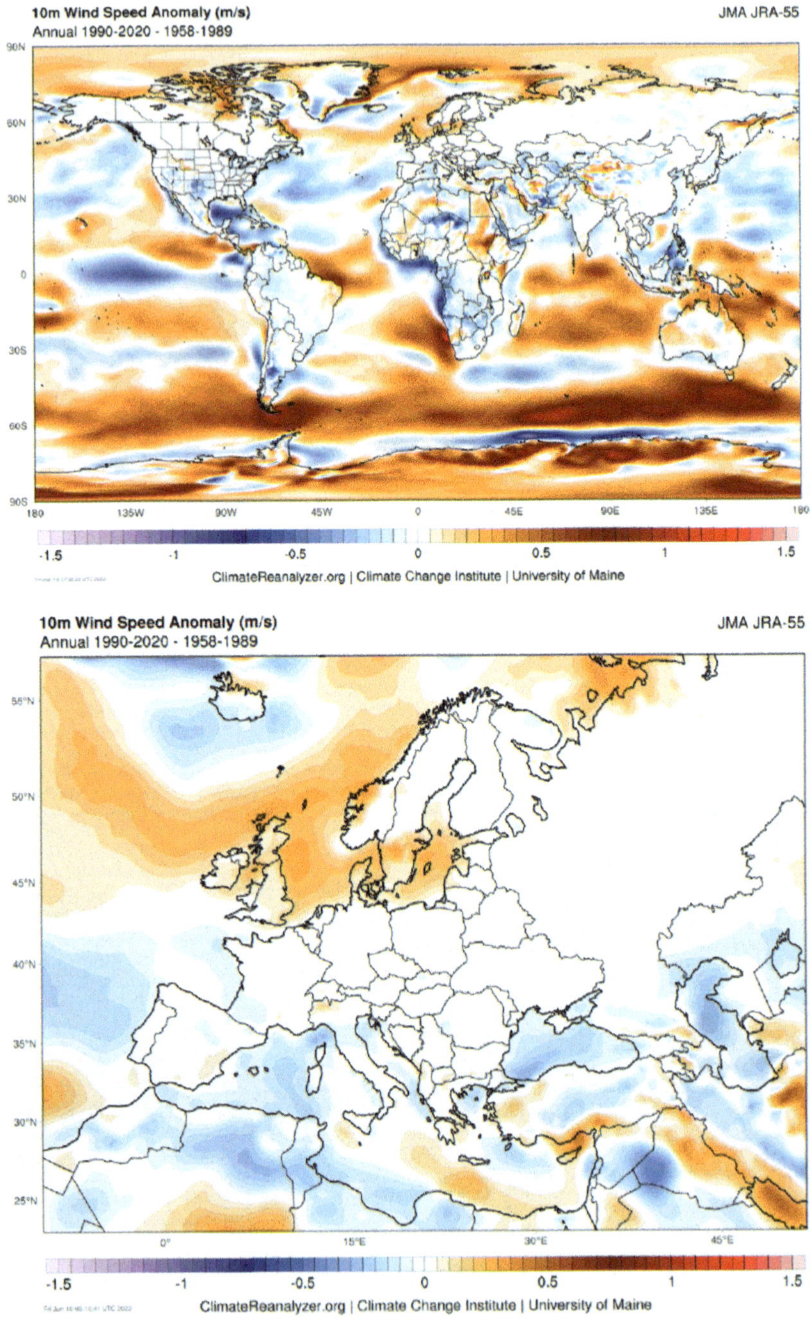


**Fig. 1.4** The juxtaposition of the mean wind speeds at 10 m above the ground in two time periods, 1990–2020 and 1958–1989. The JMA/JRA-55 analysis model (*Source* Climate Reanalyzer)

#### 1.1.2.1 Global Climate: Future Projections

The IPCC Assessment Reports evaluate the state of the climate and, incorporating the data from the past in the climate models, make projections for the future. Predictions of the severe weather situations in the next 100 years suggest that frost weather will virtually disappear [7]. Regarding Europe, IPCC warns that the temperatures will continue to rise, and the weather extremes will be more frequent and intense in the future [10]. The mean winter temperature will increase in Northern Europe compared to Central and Mediterranean Europe, while the mean summer temperature will rise more in Central and Mediterranean Europe. The mean annual precipitation will presumably increase in Northern and Central Europe. The latest sixth assessment report was released in April 2022, and almost half of the experts contributing to creating this IPCC assessment come from developing countries. One of the 6th assessment special reports [8] dealing with the impacts of increased global mean surface temperature (GMST) of 1.5 and 2 °C compared to the pre-industrial period, summarizes, with high confidence, the following consequences of such a pace of warming:

- irreversible loss of some ecosystems,
- robust increases in temperature means and extremes,
- decreases in the occurrence of cold extremes,
- extreme drought,
- heavy precipitation.

Henceforward, the target should be to stay under this limit. To avoid the rapid rise in the frequency of extreme temperatures, an unprecedented global reduction of greenhouse gases is required [2]. Projections by SMHI's (Swedish Meteorological and Hydrological Institute) climate research unit, Rossby Centre, confirm similar global future scenarios for extreme temperatures. Overshooting temperatures will occur more often and be more intense than in the past. Furthermore, truly cold days are less and less likely in the future. The scenarios by Rossby Centre also show that extreme rainfall could rise by up to 40% in Scandinavia and North-Eastern Europe [11].

A question of an increase or no significant change in average and extreme near-surface wind speeds in the future remains without a definitive answer. On the one hand, the confidence in future changes in windiness in Europe is relatively low [7, 11]. On the other hand, Eichelberger et al. [12], Nikulin et al. [13], and Forzieri et al. [14] conclude that increased annual mean wind speed can be expected in Central and Northern Europe. In the most severe climate change scenarios, the Danish Meteorological Institute predicts a rise in the wind velocity of about 5% in the next 100 years [15].

The Fourth IPCC Assessment Report reasons these contradictory predictions for central and northern Europe are model-dependent. Figure 1.5 depicts the difference in the mean wind speeds for the years 1961–1990 and the years 2071–2100. The simulation retrieved from the IPCC was created within the PRUDENCE (The Prediction of Regional scenarios and Uncertainties for Defining European Climate change

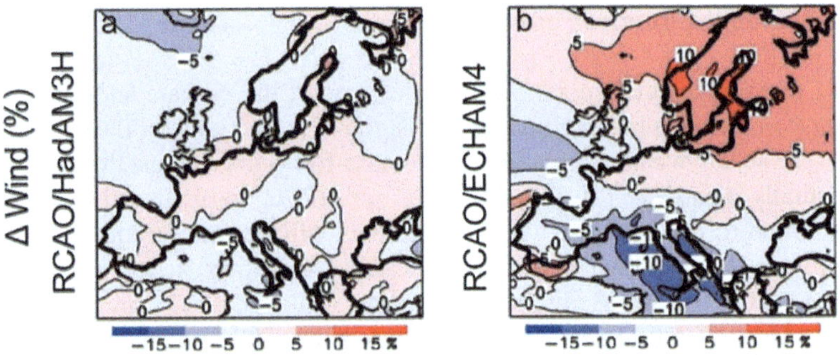


**Fig. 1.5** The fourth IPCC assessment report wind predictions (*Source* [7]—Fig. 11.6)

risks and Effects) research. When an increased north–south pressure gradient across northern Europe was used in the RCAO (Rossby Centre regional Atmosphere–Ocean model) simulation, this led to stronger winds in this area (Fig. 1.5b). Figure 1.5a, on the other hand, shows the simulation with little change in the pressure pattern and hence only minimal changes in the mean wind speed. The simulations produced employing MMD (Multi-Model Dataset) suggest that the most probable scenario will lie somewhere in-between. The occurrence of extreme wind speeds (mostly winter cyclones) in Europe is only indirectly related to the mean wind speeds, although some similarities can be observed [7].

Climate change, especially so hardly-predictable weather parameter as is the wind, has increased the need for the use of chaos theories [2] that offer exact mathematical solutions for a chaotic, non-trendy change.

### *1.1.3 Extreme Weather*

Despite the weather extremes are generally rare, the post-industrial actions of humans have led to global warming and the associated more frequent and intense extremes in weather. The nature of extreme weather events is complex, as it usually is an accumulation of more attributes. Stephenson [16] and Cutter et al. [17] define extreme events as follows:

- Extremes of a single meteorological variable: one weather variable is above or below the threshold value. The values are on the tail of the probability distribution for a location, region, or time of the year (e.g., extreme precipitation, extreme winds, such as cyclones).
- Compound (multivariable) extremes: (a) two or more extreme events occurring simultaneously or successively or (b) combinations of extreme events with underlying conditions that amplify the impact of the events or (c) combinations of events that are not themselves extremes but lead to an extreme event (e.g., a severe ice storm can involve extreme temperature, wind, and precipitation).

- Weather phenomenon associated with extremes: the weather phenomenon itself does not have to be extreme (e.g., a sandstorm formed in the arid region when strong winds blow loose sand).

### *1.1.4 Designing for Climate Change and Extreme Weather*

In the era of changing climate, it is natural that architects and engineers strive to develop environment-based, time and cost-effective design strategies. With the conventional approach to design debated, the dynamic environment is observed more closely and becomes a part of the design process [18–20]. Furthermore, the effects of the environment, even in its extreme forms, can be digitally simulated already during the early conceptual phase of architectural design to help mitigate extreme-climate scenarios [21, 22].

Digital designing, intertwining the architectural idea with environmental engineering or structural analysis, offers uncharted strategies for creating architecture. "For architectural design, it gives new tools to create buildings that use natural forces and even react interactively with the conditions, thus offering an alternative to conventional architectural design and air-conditioning practices" [2]. Edwards [23], claiming that technology is a key to sustainable design, proposes design strategies for climate change:

- The design should be focused on a building shell, its orientation, and its footprint. These factors are fundamental for building adaptability and energy efficiency.
- The buildings should have a higher initial standard, including materials.
- The designs should provide the means to upgrade building systems, especially in the areas of cooling and the provision of renewable energy.

### *1.1.5 Summary*

As a form of adaptation to climate change, climate-conscious design, along with substantial and sustained reductions in greenhouse gas emissions, can limit the consequences of ongoing climate change [6].

Climate-conscious design strategies stress the influence of climatic factors in planning, proposing to include solar irradiance or aerodynamic studies, ecology, and nature in general, in the architecture creation process.

## 1.2 The Wind

> The wind conditions of one or two simple-shaped buildings can become very intricate. Therefore, the situation with buildings of complex shapes interacting with the wind flow

**Table 1.1** Attributes of air at different temperatures (*Source* http://theengineeringmindset.com/properties-of-air-at-atmospheric-pressure)

| | −20 °C | −10 °C | 0 °C | 10 °C | 20 °C | 30 °C |
|---|---|---|---|---|---|---|
| Density[1] $\rho$ [kg/m$^3$] | 1.3958 | 1.3426 | 1.2933 | 1.2474 | 1.2047 | 1.1649 |
| Dynamic viscosity $\mu$ [$10^{-5}$ kg/(m s)] | 1.6222 | 1.6731 | 1.7231 | 1.7722 | 1.8205 | 1.8680 |
| Kinematic viscosity[2] $\nu$ [$10^{-5}$ m$^2$/s] | 1.1622 | 1.2462 | 1.3324 | 1.4207 | 1.5111 | 1.6036 |

while considering the effect of ground topography and all adjacent buildings becomes even more tricky [24].

Nature is constantly affected and, indeed, shaped by natural forces. As an analogy to the wind erosion, creating unexpected rock or sand formations in the natural environment [25], the wind in the human-made environment 'shapes' the buildings and, reciprocally, the wind flow pattern, the formation of turbulences or the wind acceleration, interrelate with the built scenery [26, 27]. A thorough digital analysis of the specific wind situation of the design site helps estimate the performance of the architectural or urban design in the wind and contributes to creating a sustainable and comfortable built environment [28].

### 1.2.1 General Properties of Air

The attributes of air at the atmospheric pressure vary at different temperatures, which indispensably impacts the character of the wind flow (Table 1.1). Throughout this book, the air properties at 20° Celsius and the values recommended by Eurocode 1 are used and integrated into the wind-related design [29].

### 1.2.2 Wind Direction and Magnitude

Due to the alternating character of the wind direction, the meteorological observation points on Earth have determined the prevailing winds for each observation site. The prevailing winds are a crucial indicator of the wind microclimate of a given place. Figure 1.6 depicts a simplified analysis of the global average wind directions along the coastlines and inland, representing the data over the last 20 years, measured 10 m above the surface. The records are retrieved from the CCI—Reanalyzer [9]. The representative wind flowlines are drawn on the Fuller projection world map illustrating the average wind speeds and directions. The dotted blue curves in Fig. 1.6

---

[1] The recommended value by the EN 1991–1-4:2005 for the calculations of wind load is 1.25 kg/m$^3$.

[2] The recommended value by the EN 1991–1-4:2005 for the calculations of the wind load is $1.5 \times 10^{-5}$ m$^2$/s yy[29].

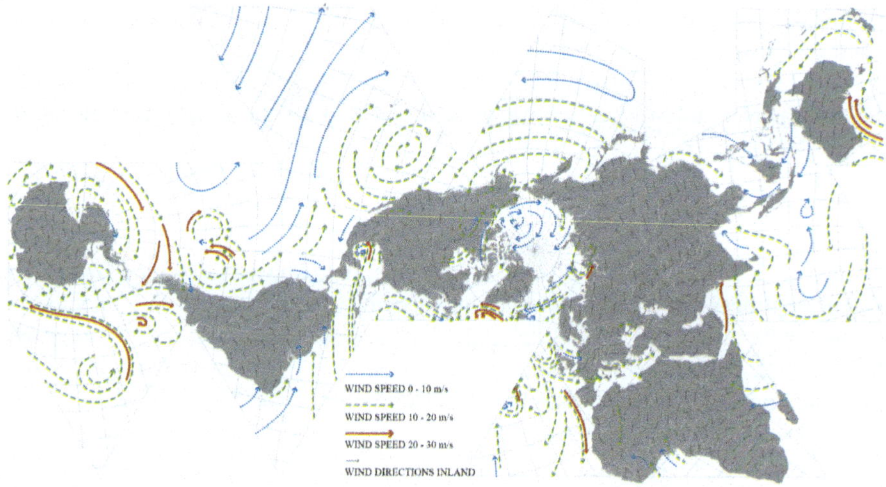


**Fig. 1.6** Global average wind directions and speeds (*Data source* CCI-Reanalyzer, map source: The Buckminster Fuller Institute)

represent the wind speeds between 0 and 10 m/s. The dashed green curves show the wind speeds between 10 and 20 m/s, and the thick red continuous curves represent extreme wind speed, reaching 30 m/s.

### *1.2.3 The Eurocode*

The Eurocode 1, EN 1991-1-4:2005 *Actions on structures*, initially published in 1991 [29], provides regulations and guidance for the structural design of buildings, which are up to 200 m high, and civil engineering works, which are up to 200 m in span, loaded by the wind. It is used in nearly all EU member states, excluding Bulgaria, Croatia, and Romania. Conversely, Iceland, Norway, and Switzerland follow Eurocode 1 in the design of building and engineering structures affected by the wind. The core of the Eurocode is inscribed in all national versions. When alternative options are possible for values, procedures to be used, or country-specific data (e.g., wind map), the choice is country-specific, incorporated in forewords and annexes. There are efforts to merge the country-specific fundamental values of the Basic wind velocity $v_{b0}$[3] into one consistent wind map for Europe [15]. This value represents extreme wind speeds and characterizes the wind conditions of different areas in Europe. The influence of climate change on the increasing wind speed has

[3] $V_{b0}$ is the characteristic 10 min-mean wind velocity with an annual exceedance risk of 0.02 (equivalent to the mean return period of 50 years), irrespective of wind direction and season, at the height of 10 m above flat open country terrain category II (see Table 1.2).

not been coherently considered in the wind maps. However, the effects of increased wind speeds are to be incorporated into the European design standards [15].

The Basic wind velocity relates to the Mean wind velocity $\boldsymbol{v_m}$, representing the 10-min-mean wind velocity considering the effects of the terrain roughness and orography:

$$v_m(z) = c_r(z)c_0(z)v_b \tag{1.1}$$

where $v_m(z)$ is the mean wind velocity at the height $z$ above the ground [m/s]; $c_r(z)$ is the roughness factor, representing the variability of the mean wind velocity due to the height above the ground upwind of the structure [−]; $c_0(z)$ is the orography factor (=1.0, unless specified otherwise) [−]; $v_b$ is the basic wind velocity, defined as a function of the wind direction,and the time of the year, 10 m above the ground of terrain category II (see Table 1.2).

## *1.2.4 The Wind and the Built Environment*

Caused by the inconsistent heating of the Earth's surface, the air moves from a high- to a low-pressure area, creating the wind flow. Obstacles to wind flow can generate pressure differences; a high-pressure area is formed on the windward side, while on the leeward side, a low-pressure area emerges [2]. The resultant wind speed, pressure, and flow lines in the urban environment are influenced by the interplay of several aspects, such as:

- the configuration and urban design of buildings,
- terrain morphology and roughness,
- building shape.

### 1.2.4.1 The Configuration and Urban Design of Buildings

As was already implied, the built environment, consisting of the positive (matter) and negative (voids) space, contributes to "shaping" the natural air movement [30]. That denotes a commitment while creating/designing new urban areas or buildings, especially ones located within the context of existing buildings. While there are no rules or guidelines for designing in relation to the wind microclimate, Eurocode 1 does designate several regulations for the design of buildings neighboring high-rise buildings:

- an increase in wind velocity for some wind directions could affect the planned building if it is close to a high-rise building, which is at least twice as high as the average height of neighboring ones.
- the increased wind velocities can be disregarded in the design process if the height of the proposed building is more than half the height of the high-rise building.

**Table 1.2** Terrain roughness categories

| Terrain category | $z_0$ [m] | $z_{min}$ [m] |
|---|---|---|
| 0 = sea; coastal area, exposed to the open sea | 0.003 | 1 |
| I = lakes; areas with negligible vegetation and without obstacles | 0.01 | 1 |
| II = areas with low vegetation (grass) and isolated obstacles (trees and/or buildings) separated by > 20 × obstacle height | 0.05 | 2 |
| III = areas with a regular vegetation cover; areas with buildings or isolated obstacles with separation < 20 × obstacle height (e.g., villages, suburban terrain, permanent forest) | 0.3 | 5 |
| IV = areas where buildings cover > 15% of the surface, while their average height > 15 m | 1 | 10 |

#### 1.2.4.2 Terrain Morphology and Roughness

In a non-mountainous region, an isolated hill (ridge, cliff, or escarpment) causes an increase in the wind velocity with the peak near the top of the slope. If the wind velocity increases by more than 5%, the orography factor $\boldsymbol{c_0}$ should be employed in the wind analysis and used in the mean wind velocity $\boldsymbol{v_m}$ calculations (see Eq. 1.1). These orography impacts may be neglected if the upwind terrain (the windward side) has an average slope of less than 3°.

Furthermore, the wind speed is influenced by the terrain roughness, expressed in calculations by the terrain roughness parameters $z_0$(roughness length) and $z_{min}$ (minimum height). Due to the friction, the wind speed decreases closer to the terrain. The decrease depends on the terrain category (Table 1.2) [29]. This terrain roughness effect is observed to a certain height (300–500 m) above the surface, depending on the terrain category, and the wind speed is considered constant above this height. In the wind-driven design introduced in this book, the wind speed 10 m above the ground will be used, as the wind data from meteorological stations are measured for this specific height.

#### 1.2.4.3 Building Shape

The overall wind situation in the built environment, the pedestrian wind comfort, as well as the wind pressure acting on buildings are all influenced by the size, global and local building shape, as well as dynamic properties of structures [27, 31, 32]. Public spaces where the urban configuration and the shape of buildings cause wind acceleration or turbulence are avoided by people. That disrupts outdoor activities and, in addition, can cause wind-related accidents [33].

For this reason, wind speed threshold values and the exceedance percentage are designated and utilized in pedestrian wind comfort evaluation. If the wind speed exceeds the comfort limit of 5 m/s, set by the Dutch standard [34] >10% of the time, this results in poor conditions for sitting and strolling activities. The Lawson criteria [35] specify a higher threshold level; wind speeds from 5.3 m/s represent the onset of discomfort and are not suitable for sitting or a slow walk. If the wind speed is higher than 5 m/s > 20% of the time, a space is considered uncomfortable for any activity [24, 34]. Wind speeds exceeding 15 m/s > 0.05–0.3% of the year can cause wind-related danger in the public space.

## 1.2.5 The Wind Loads

Wind actions change with time, acting directly as pressures and forces on the enclosed building envelopes. Because of the porosity of the outer surfaces, the wind indirectly influences the inner surfaces too. The building envelope of open structures, on the other hand, is directly exposed to wind actions [29]. The wind loads on structures result in their dynamic and static responses. Dynamic properties of the wind can be neglected if the stiffness of the building is sufficient. Then, the wind load acting on such a structure is static and evenly distributed along the surface. The wind loads act perpendicularly as positive (+) or negative (−) pressure, ergo suction. Friction acts parallel to the building surface. The pressure coefficients, as well as force coefficients, express the wind effects on buildings and are known as aerodynamic coefficients.

### 1.2.5.1 The Wind Pressures

The wind pressure at any given point of the wind flow at a specific moment can be expressed by Bernoulli's equation [36]:

$$p = p_{\mathrm{a}} + \frac{1}{2}\rho v^2 \tag{1.2}$$

where $p$ is the wind pressure [Pa], $p_{\mathrm{a}}$ is the atmospheric pressure (static pressure) [Pa], $\rho$ is air density [kg/m$^3$], $v$ is the wind speed [m/s]. The expression $1/2\rho v^2$ represents kinetic energy (dynamic pressure).

After impacting a building, the wind changes its direction and decreases the velocity from the initial velocity $v$ to the velocity $v_1 = 0$, which manifests as the wind surface pressure on the obstacle.

#### Pressure Coefficients

The pressure coefficients represent the pressure effects of the wind on buildings. On the external surfaces, these effects are demonstrated through the external pressure coefficient $\boldsymbol{c}_{\mathbf{pe}}$, while on the internal surfaces of buildings as internal pressure coefficient $\boldsymbol{c}_{\mathbf{pi}}$. The general pressure coefficient $\boldsymbol{c}_{\mathbf{p}}$ can be defined as a difference between the external and internal pressure coefficient.

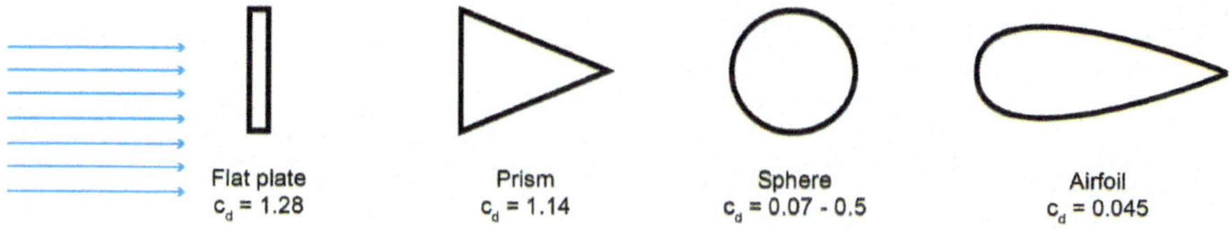


**Fig. 1.7** The effect of the shape on drag coefficient

Force Coefficients

Force coefficients give the overall effect of the wind on a structure, structural element, or component, including friction, if not specifically excluded [29]. When the wind force acts on a body (in our case buildings) in the upwind direction of the incident wind, the drag coefficient $\boldsymbol{c}_{\mathbf{d}}$ can be determined as follows:

$$c_{\mathrm{d}} = \frac{2F_{\mathrm{d}}}{\rho A v_m^2} \tag{1.3}$$

where $c_{\mathrm{d}}$ is the drag coefficient [−], $F_{\mathrm{d}}$ is the drag force [N], $\rho$ is the air density [kg/m$^3$], $A$ is the reference area of the object exposed to the wind [m$^2$], $v_m$ is the mean wind velocity [m/s] at a building height $h$ in the upstream, undisturbed flow.

The drag coefficient and the drag force are dependent on the obstacle's shape and the resistance to the flow. With a lower drag coefficient, the aerodynamic tension in the wind decreases [37]. The shape optimization in the conceptual design stage is, therefore, beneficial not only in the automotive or aerospace industry but also in urban and architectural design, with the objective of lower surface pressure and hence lighter building structures. Experimentally obtained drag coefficient values from the wind tunnel measurements of various shapes with the same frontal area are shown in Fig. 1.7 [38].

With the wind force perpendicular to the direction of the incident wind, the resulting force is called lift, defined through the lift coefficient $\boldsymbol{c}_{\mathbf{l}}$. The calculation is, in principle, the same as in the case of the drag coefficient.

### *1.2.6 Summary*

The reciprocal influences of the wind and building shape are closely associated with the role of the wind in urbanism. In the wind-driven architectural form-finding, the basic properties of air, the prevailing direction of the wind flow, and the wind magnitude can be incorporated as inputs, influencing the final urban/architectural/building envelope design.

On the urban scale, a cluster of buildings jointly affects the wind flow resulting in complex mutual relations regarding the wind. A single building within a group impacted by the wind flow might adversely influence the rest of the buildings, the wind flow pattern, and the wind loads.

## 1.3 Algorithms-Aided Design and Simulations in Architecture

Simulating natural phenomena like internal and external airflow is gaining increasing significance in architectural design [39].

Architectural design is an intricate process, with different inputs and criteria of varying significance entering the design decisions and consequently playing a role in the final design. Over the past decades, computers have become an essential part of everyday life. With enhanced hardware and digital software, it is only natural that digitalization is engaged in the architectural design process. Algorithms-Aided Design enables entwining the ingenuity and the knowledge of the designer with the capacities of computers and exploits both in the process of architectural designing.

### *1.3.1 Definitions*

- **Parametric designing**. Geometric constraints-based form generation [40].
- **Form-finding (from the structural engineering angle)**. Finding a shape of equilibrium of forces in a given boundary [41].
- **Form-finding (from the architectural design angle)**. Employing digital design tools, finding an appropriate architectural and structural shape, based on the criteria such as material properties, geometric behavior, manufacturing requirements, or environmental characteristics [41]. Form-finding engaged in architectural design enables the creation of complex and unexpected forms [42].
- **CFD (Computational Fluid Dynamics)**. A branch of fluid mechanics that deals with numerical simulation of fluid flow based on the conservation laws (conservation of mass, momentum, and energy, calculated from Navier-Stokes equations) governing fluid motion and complex flow phenomena [43, 44].

### *1.3.2 Algorithms-Aided Design*

Parameter-based algorithmic modeling has yielded greater freedom in designing complex architectural shapes and is a fundamental player in creating balanced technology-architecture-nature relations that result in environment-responsive designs. Through parametric modeling, the designer can integrate the environmental (and other) aspects and iteratively evaluate the design's performance while changing the input criteria and design goals [26, 45–47].

Parametric designing streamlines the process of creating the most suitable design option fulfilling specific constraints [48]. Utilizing Rhinoceros[4] and its algorithmic plug-in, Grasshopper[5] enables exchanging the information between the design and simulation software, thus creating several environment-responsive design choices [49]. Leveraging open-source software might lead to a true digital revolution in the architectural design practice [50].

## *1.3.3 CFD Simulations in Architectural Design*

CFD is a modern computational method primarily used in the aeronautics and aerospace industry. It has been gradually integrated in the automotive, chemical, nuclear, and marine industries. With the rapid improvements in hardware and CFD software over the last decades, the idea of employing the CFD technique in the architectural design process was just a matter of time. An initiative of an online CFD teaching project was visible as early as 1998 to introduce the novel method to architects [30]. Still, around the new millennium, CFD was a rather new technique in building design, used solely by indoor climate specialists, while almost unknown to architects. Nowadays, even though wind flow simulations are advantageous in architectural design [51], it is still not typical to leverage wind analysis in the early stage of architectural design [52]. Admittedly, this method is well integrated into the civil engineering area for heating simulations, ventilating and air-conditioning simulations, fire simulations, fluid–structure interactions, or air quality assessment [44]. CFD utilized in architectural design enables transferring the complex relations of the wind into the architectural context, analyzing their effects, and, most importantly, exploiting the benefits of the air movement in the built environment [27, 53].

The global wind flow has uneven direction and speed (see Fig. 1.6). On a smaller scale as well, the wind is tightly intertwined with the specific location on Earth, creating a particular microclimate with a typical wind situation. For this reason, the wind-driven design, fusing CFD simulations with architectural form-finding, can be aimed at different goals [54].

### 1.3.3.1 Natural Ventilation

In hot climates, knowing the effects of the wind flow as early as in the conceptual design phase can be leveraged to achieve natural ventilation between buildings [55]. The reciprocal rotation of buildings, the size of the public space between them, as well as the height of the buildings affect the character of the wind flow.

---

[4] https://www.rhino3d.com/.

[5] https://www.grasshopper3d.com/.

The research of Taleb and Musleh shows that up to a 30% enhancement in natural ventilation can be attained through incorporating the weather information into the design, employing genetic algorithms in the design process and thus generating multiple design options, and finally evaluating their performance by exploiting CFD simulations [47]. Existing urban areas can also be investigated, although it is much harder and costlier to make changes or adjustments that would improve the ventilation rate [56].

Not only can the potential of natural ventilation be put to use on the urban scale [57, 58], but also on an architectural scale, for example, through meticulous designing of double, permeable building envelopes [19, 32]. Further, by employing CFD simulations in the design process, an interior shape-optimization can be performed to create the optimum shape of the building's interior in terms of natural ventilation [59].

#### 1.3.3.2 Pollutant Dispersion

Urban spatial characteristics, including the site coverage ratio, buildings' frontal area density, height, and height diversity, influence the pollutant dispersion in the densely-populated areas [60]. The research of Zhang et al. specifies that the urban air quality (the concentration of almost all air pollutants) mainly depends on the 2D urban landscape patterns, ergo the distribution of block and building footprints, while the 3D urban form shows insignificant effects. Moreover, the built landscapes affect the air quality more than the natural morphology [61]. Early-stage designing with CFD could predict the extent to which the designed solution affects dispersing traffic pollution in big cities.

#### 1.3.3.3 Aerodynamic Architectural Forms

Streamlined building forms designed in colder or very windy climates can lead to a reduction of the wind loads [1]. Wind-based structural optimization can effectively reduce the material cost of supertall buildings, even when dealing with a complex structural system [62]. Also, altering the facades of existing buildings with minimal architectural interventions contributes to reducing wind loads on building surfaces [63]. He overall building shape and the roughness of its façade impact the wind flow and, consequently, the pressure acting on the façade [64]. Moreover, the wind-driven design is invaluable when the building structures are subjected to high-speed winds, such as in hurricanes. CFD is a principal virtual tool assisting in mitigating wind damage to buildings [65].

#### 1.3.3.4 Pedestrian Wind Comfort

For pedestrians, the negative impacts of the wind are perceived through the level of wind comfort outdoors and indoors. Systematically planned urbanism and suitable architectural intervention can regulate or even reverse wind-related discomfort [66]. Poćwierz and Zielonko-Jung have demonstrated that a linear arrangement of houses and medium land-use efficiency result in well-ventilated zones, whereas the wind velocity (in standard weather conditions) does not exceed threshold values for pedestrian wind comfort [67]. The wind can be accelerated in-between buildings, resulting in an uncomfortable pedestrian passage between them. A well-designed wind barrier can deflect the airflow further behind the obstacle while acting as a decelerator of the flow resulting in enhanced wind comfort at the street level [68, 69]. In addition to the decelerating effect, placing a porous barrier in the regions with severe sandstorms can reduce the number of sand particles downwind the obstacle [70].

#### 1.3.3.5 Harvesting of the Wind Energy

The building shape [71] and the urban configuration of buildings [72] influence wind speed and, consequently, energy harvesting effectivity. Massive wind turbines mounted on buildings are one way to generate electricity. Piezoelectric[6] materials applied to facades can serve the same purpose. Their efficiency, however, compared to the classical genera4ors, is much lower, ergo, 20–70%, compared to 0.5–15% for piezoelectric and classical generators, respectively. Yet design-wise, they undoubtedly are a promising way of combining wind energy harvesting with architecture [27, 73]. An intriguing thought is a dynamic building façade composed of foldable units, which harvest wind energy and generate electricity by employing a system of cables and a generator rotor [74].

### *1.3.4 CFD Analysis*

#### 1.3.4.1 The Virtual Model Setup

A typical fluid dynamics problem involves basic fluid properties like flow velocity, pressure, density, and temperature, considering time and space. The CFD simulation goal is to determine the mean flow with acceptable precision. A virtual (digital) 3D model of the tested object(s) is a prerequisite to the correct simulation setup. Further, the initial conditions, as well as boundary conditions, are determined, and the wind tunnel is defined. The simulation results more accurately represent the real-life situation in case the correct dimensions are designated for the virtual wind

[6] Piezoelectricity is a property that allows materials to convert mechanical energy to electrical energy and, conversely, electrical energy to mechanical energy.

tunnel. The wind flow observed around the tested object is thus not disturbed by the walls of the wind tunnel. If an $h$ is the height of the highest building in a problem case, the best CFD guidelines advise using the length of the upstream zone as 5x$h$, the length of the downstream zone 15x$h$, the height 5x$h$, and the width of the domain 5x$h$ from each side of the investigated case [75, 76]. Simultaneously, the buildings' projection in the domain's transversal section should not exceed 3% of the section (so-called domain blockage ratio), although, for simulations at the urbanism scale, these restrictions might be too strict [77].

CFD simulations are useful for parametric studies of architecture exposed to the wind flow, where multiple design alternatives are considered. Such wind analysis would be impractical or impossible if performed theoretically or experimentally. Many projects have proved the usefulness of parametric studies intertwined with CFD simulations [27, 45, 47, 58]. Even though parametric design coupled with CFD is promising, it still has a few drawbacks:

- **The interpretation of the CFD simulation results of multiple design options merely through visual comparison might be problematic**. Not all the nuances in the results are detected by the designer/engineer. That, however, could be resolved by a computer script, which would distinguish the differences [78]. Another way for easier interpretation of the CFD results might be employing Virtual Reality (VR) in the post-processing of the results. This approach would blend the computer capacities with the designer's experiences and interventions [55, 79].
- **If multiple design options are considered in the design process, the necessity of repetitive CFD solving for each investigated option is a drawback** [80]. The performance of the designs in the wind is evaluated in a loop. The architecture/urban plan is designed, then a CFD tool is utilized to determine how the design affects the wind, and finally, the CFD results are visualized and post-processed. Once multiple design options are tested, the loop repeats until the best-performing solution is found. Every single design change, adjustment of the geometry, or anything influencing the CFD simulation (including the initial wind speed or boundary conditions), requires the designer to remodel the geometry, re-mesh it, and finally run the simulation to recompute the airflow [45]. However, the designer can overlook some fitting solutions [49]. Evolutionary algorithms combined with visualization can help with this difficulty, offering an automated shape optimization [81]. The disadvantage is the limited accuracy and the fact that the automated optimization might interfere with designers' deliberate, wind-driven design decisions.
- **Accurate CFD simulations require longer computing time**. Simulations of large-scale urban areas are the most time-consuming, requiring a powerful computer. An urban zone 1 $km^2$ simulation case can consist of as many as 10 million computational cells [80]. There are potential solutions to this issue too, including the Fast Fluid Dynamics (FFD) approach, Machine learning (ML) approach, and a combination of the two. This will be discussed in more detail in the following section.

**Table 1.3** Comparison of the wind analysis tools (+ advantages, − disadvantages)

| | Rhino CFD[7] (for Rhinoceros) | Swift/procedural compute[8] (for Grasshopper/Blender/Revit) | Butterfly[9] (for Grasshopper) | InFraRed (web-based) |
|---|---|---|---|---|
| Availability | + Developed in 2018 + Available as a plug-in for Rhino + Free license for students | + Developed in 2018 + Now available as a cloud-based tool − The cloud version is paid | + Developed in 2017 + Available as a Grasshopper plug-in + Free | + Developed in 2020 + Available online (http://infrared.city/) + Free |
| Virtual Wind Tunnel (VWT) settings | + The domain and its relative position to the tested geometry can be precisely specified | − The domain and its relative position to the tested geometry are specified as multiplications and offsets | + The domain and its relative position to the tested geometry can be precisely specified | + The analysis domain is specified without the need for the VWT |
| Calculation speed | − Many settings must be addressed before the start of the simulation − The calculation time is standard | + The calculations run on the cloud − The calculation time is standard (runs on the OpenFOAM platform) | − The calculation time is standard (runs on the OpenFOAM platform) | + The 2D wind comfort/speed prediction is available in a matter of seconds |

#### 1.3.4.2 CFD Tools

Nowadays, many suitable CFD (or other wind analysis) software options are accessible to architects and planners [83]. Table 1.3 summarizes the advantages and limitations of four of them [81, 84, 85].

The Grasshopper CFD plug-ins discussed in Table 1.3 use OpenFOAM, a well-established and accurate CFD engine [83] that can be controlled and run through the

[7] https://www.cham.co.uk/rhinoCFD.php.

[8] https://compute.procedural.build/.

[9] https://www.ladybug.tools/butterfly.html.

Grasshopper environment. For the post-processing of the results, however, a standalone software Paraview[10] is needed to visualize the simulation results and retrieve various outputs, including calculated variables (e.g., wind surface pressure). The necessity of the external post-processing software after the CFD analysis may seem like a disadvantage. However, there are tricks for fast results preview for Butterfly and Swift/ Procedural Compute. Moreover, Paraview can process a large amount of data fluently thanks to high-performance computation techniques. Rhino CFD, on the other hand, enables displaying the results directly in Rhino, whereas InFraRed returns the instant wind prediction/comfort image.

#### 1.3.4.3 Other Wind Analysis Techniques

**Fast Fluid Dynamics**. That is another approach to analyzing the wind situation around architectural objects. It is faster than standard CFD analysis but has lower precision in capturing the wind flow correctly [83, 86]. FFD was developed in 1999, specifically for computer visualizations in video games. Its implementation in wind simulation around buildings is of a recent date. The method is not yet widely used, as the FFD results compared to the OpenFOAM results show many discrepancies between the two, probably because of the lack of spatially resolved turbulent effects [87]. This type of wind analysis fits for, e.g., fast preliminary urban design tests.

**Machine learning approach**. Trained on the thousands of CFD simulation cases, the machine learning-based tool InFraRed, developed by the CIL of the AIT in Vienna,[11] is an algorithm predicting the wind flow [88, 89]. The algorithm can provide a 2D wind comfort and wind flow prediction at the pedestrian level in a matter of seconds for areas sized 250 × 250 m. Larger zones can be analyzed, reaching an accuracy suitable for the first design stages [90].

**A hybrid method combining FFD results with ML model**. This experimental approach uses the FFD results and feeds them as inputs to the ML model. This way, training the model does not require much time, the fidelity of both approaches is improved, and the simulation time remains low. The hybrid approach might even overcome classical ML models [91].

#### 1.3.4.4 CFD Versus Physical Wind Tunnel Tests

Wind analysis-oriented performative design, incorporating CFD simulations, is usually carried out in the final design stages when the alterations to architectural designs are infeasible and sometimes impossible. In contrast, employing the CFD simulations in the early design stage offers a flexible, fast, and interactive setting for studying the performance of a variety of design concepts in the wind [30, 58, 90]. In the later design stages, once only one or two best-performing alternatives are

---

[10] https://www.paraview.org/.

[11] City Intelligence Lab of the Austrian Institute of Technology in Vienna, Austria.

**Table 1.4** The comparison of advantages (+) and disadvantages (−) of using a wind tunnel and CFD analysis

| Wind tunnel | CFD |
|---|---|
| − expensive | +cheaper |
| − time-consuming | +faster |
| − investigating one quantity at a time | +simultaneous computation of instantaneous density, velocity, pressure, temperature, and concentration of fields |
| − problematic experimentation with extreme pressures and extreme temperatures | +arbitrary extreme conditions can be applied |
| − a small-scale model must be built, although, +the scaled model can be 3D-printed | +full-scale solution |
| − testing various design options is time-consuming | +parametric studies |
| − measurements errors | − 3D models can cause errors |
| − flow disturbance caused by probes | − input parameters can be set incorrectly which influences the results |
| +the wind flow can be observed in the reality | − the wind flow is observed on a computer screen (except for the VR results display) |

selected, wind tunnel testing can be utilized for further investigation. Employing the wind tunnel during the early design phase is neither cost-effective nor convenient for making crucial design decisions [30, 92]. With the increasing computing power, CFD simulations are faster and easier to perform. The comparison of the results obtained through the CFD simulations with the wind tunnel measurements was performed by many researchers, validating this digital methodology against the physical wind tunnel tests [43, 76, 93] or on-site measurements [75]. Table 1.4 depicts the main benefits and drawbacks of CFD and wind tunnels.

### *1.3.5 Summary*

CFD wind analysis certainly is a good starting point for evaluating the wind-related performance of buildings and, if necessary, defining problems to be verified through the physical wind-tunnel experiments [94]. Using Grasshopper for Rhino for parametric modeling of geometry, with the subsequent use of Rhino CFD or a CFD plug-in for Grasshopper, appears to be a suitable approach for repetitive CFD testing of several design options, enabling the designer to model within Rhino and Grasshopper environment, control the design parametrically through Grasshopper, and quickly create the wind analysis within the same working environment. FFD and ML-based techniques are promising, especially for use in the early stages of design and for testing large urban zones.

## 1.4 Adaptive Architectural and Structural Solutions

> Adaptive architecture can develop many forms: Morphing theatres dancing with performers, houses that shrink to decrease their surface area and volume in winter or that close up their envelope in summer, or…buildings altering their aerodynamic profile to reduce wind loads [95].

The definition of adaptive architecture is not firmly established in architectural research and practice. There are categorizations, however, determining principles of adaptive architecture. Furthermore, examples of realized architecture and design prototypes set an innovative foundation for the potential developments in this field [96–99].

Experiencing the climate change era, architects, planners, engineers, and others might question the conventional principles of creating architecture. Therefore, the search for nature-responsive, morphing, or even 'living' buildings sets off the new frontline for architects. Entwining architecture with nature, incorporating the immaterial climatic fluxes as an integral part of the design erases the boundaries between the natural and human-made [99]. Adaptive building skins might be a way to blend technology and nature. As building skins are exposed to the effects of the weather and the dynamic needs of the occupants (Fig. 1.8), adaptive skins could conform to the changing situations and comply with the weather influences and external loads instead of passively resisting them [100]. The benefit is that the change in the shape of the building façade is only temporary. After the loads (or external inputs) no longer act, the building skin can resume its original configuration.

### *1.4.1 Adaptive Systems Principal Categorization*

The way the adaptive architecture reacts to the changing ambient conditions designates two categories of adaptive systems:

- Material-based systems. These systems adapt to the changes through the material-embedded physical properties.
- Geometry- and material-based systems. These systems leverage the advantages of material-based systems and the spatial geometry configuration, adapting to the changing conditions through the overall geometrical form change.

The purpose of architecture is to provide shelter from changing environmental conditions. In addition to that, adaptive architecture can actively cope with dynamic loading or other ambient conditions that result from environmental changes [95]. So-called 'soft' adaptive systems, which do not require sensors to 'sense' the ambient conditions nor computers to respond to stimuli, are preferred over so-called 'hard' mechanical approaches in the long-term use. The 'hard' approaches could prove unfeasible over time [101, 102].

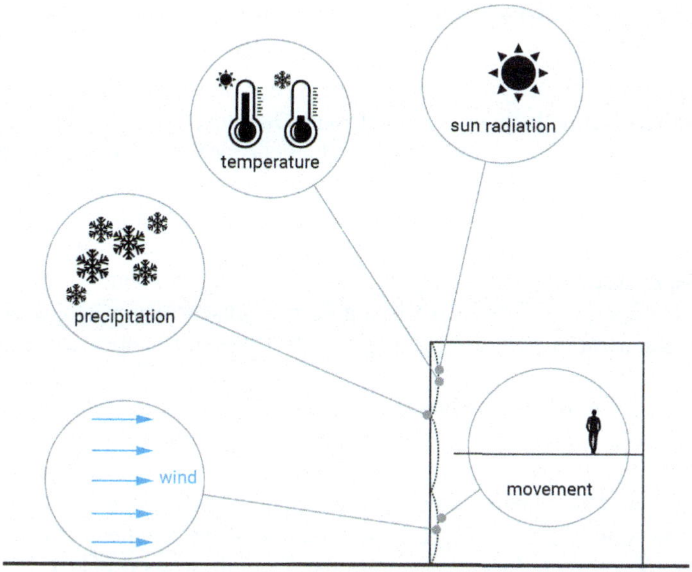


**Fig. 1.8** Various external and internal stimuli that affect buildings

#### 1.4.1.1 Material-Based Systems

Properties of some materials enable a passive change in shape without the need for complicated mechanisms and motors. Such structures, applied onto building envelopes, respond to the changing ambient conditions, such as the humidity [103, 104], heat [19], [105, 106], or mechanical load [107] by expanding or shrinking, hence, the changes at the micro-level are transformed into changes in shape at the macro level.

#### 1.4.1.2 Geometry-and Material-Based Systems

Even though the material properties can trigger off the shape reconfiguration, the overall shape change of a façade (building skin) element is facilitated through its geometric configuration. In these form-active structures, the form and the applied loads are interdependent. Examples include bending-active structures, which create 3D-curved geometries in static equilibrium from initially flat 2D geometries. The 3D geometries are created through acting force, producing: bending-active structures [108–110], or tensegrity (tensional integrity) structures [64, 111–113].

## 1.4.2 Tensegrity Structures

The tensegrity structural systems have several recognized inventors: Richard Buckminster Fuller, Kenneth Snelson, and David Georges Emmerich [114]. Tensegrities are systems in a stable (with the mechanical meaning attached to this word) self-equilibrated state, containing a discontinuous set of compressed bars (struts) that remain in equilibrium inside a continuum of tensioned cables (tendons) [111]. Within the system, all elements are interdependent, so a slight change in the potential energy of the structure based on a given applied force results in a transformation of the overall shape [101, 115]. Like living forms, tensegrities produce an optimum structural organization and are predisposed to optimum energy and material use [116].

Tensegrity systems can be enhanced to create active structures, adapting to external loads. Selected compression members (struts) can be replaced by telescopic struts, operating as active members within the system. As this option usually leads to slack tendons and significant damping is introduced, the tension in tendons must be adjusted accordingly [117, 118]. Replacing some of the tensile elements (tendons) with linear actuators produces a structure free from these issues [119, 12sw0].

When integrated into adaptive envelopes, tensegrities can either employ the pure tensegrity principle [95] or fuse it with folding [101, 121], or bending [122]. Despite the complexity of shape-altering structures, the concept of a morphing roof or a façade responding to environmental conditions is intriguing. The benefits of tensegrity structures over conventional façade and roof systems involve [101, 119, 120]:

- A change in potential energy (applied force) needed to alter the original shape of the tensegrity is minimal due to the kinematic indeterminacy.
- The tensegrity systems are deployable. From a closed configuration, they can expand to a predetermined form.
- They are material-efficient. Maximum strength can be acquired using minimum mass.
- The struts and tendons can act as sensing or actuating elements (multi-functionality).
- Structural stability is ensured by the ability to react as a whole system rather than as a part. The local stress is absorbed through the structure.
- Elasticity after the loads stop acting. Tensegrities assume the original shape after there is no external load.

#### 1.4.2.1 Digital Tools for Designing the Tensegrity Structures

The algorithmic modeling software Grasshopper was mentioned succinctly in the section Algorithms-Aided Design. Kangaroo 2[12] is an add-on for Grasshopper, which, employing the particle-spring system [123], facilitates finding an equilibrium

[12] https://www.food4rhino.com/app/kangaroo-physics.

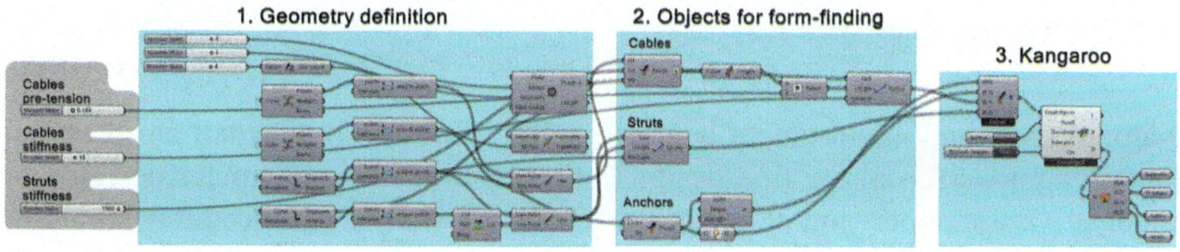


**Fig. 1.9** The Grasshopper definition depicts the consecutive steps with Kangaroo 2 solver for finding an equilibrium of one tensegrity unit

of forces. The materials are characterized through values of stiffness and pretension for the whole defined geometry (Fig. 1.9).

Using iterative calculations, the Kangaroo 2 solver finds an equilibrium state for every new value of material stiffness and pretension [101, 119]. The inputs for stiffness values are calculated from Hook's law:

$$k = \frac{EA}{l} \tag{1.4}$$

where $E$ is the Modulus of elasticity [Pa], $A$ is the cross-sectional area [$m^2$], and $l$ is the length [m].

The resulting form-found shape of a four-strut tensegrity unit in static equilibrium is schematically depicted in Fig. 1.10.

Tensegrity modules can be arrayed to create a spatial form and coat almost any surface with a continuous network of reciprocally interconnected tensegrity units. On a building envelope, the adaptive response can occur at the module level; ergo, each tensegrity module or a group of modules can morph independently within the network; or, the envelope can morph as one entity, each module triggering a movement in the neighboring one. Despite the latter option being a very complex structural system, the digital design techniques enable the designing, analyzing, and subsequent construction of these systems.

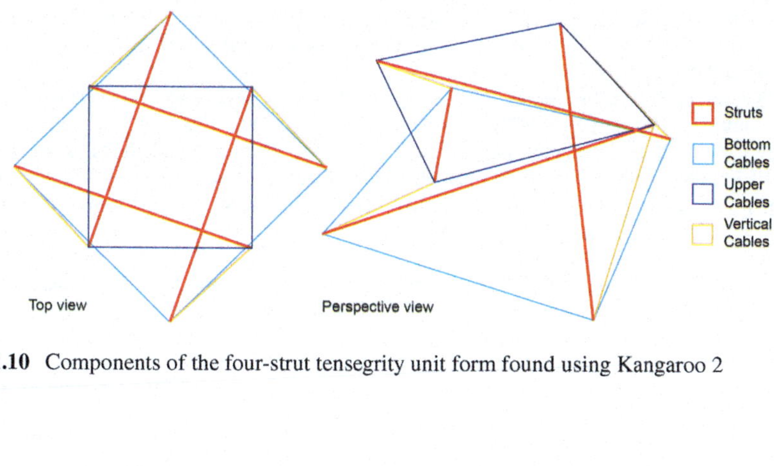


**Fig. 1.10** Components of the four-strut tensegrity unit form found using Kangaroo 2

### 1.4.3 Summary

Nature-driven and environment-adaptive architectural designs entail an interdisciplinary approach from the first design concepts. Innovative computer design tools such as Grasshopper and its plug-ins for wind analysis (Procedural Compute, Butterfly), or structural analysis (Kangaroo 2) enable architects and planners to integrate weather and structural constraints into the conceptual design phase, which, consequently, affects the outcome, whether it is architecture or large-scale urbanism.

The wind-architecture or wind-urbanism interactions within the context of any site can be predicted and leveraged during the wind-driven design.

A building envelope designed from the tensegrity structural system could reconfigure its shape with the dynamic wind force as a trigger, impacting the forces distribution within the system and consequently enabling the design of lighter structures.

Furthermore, the reciprocal interactions of a cluster of buildings with adaptive building skins must be anticipated and analyzed from the conceptual design phase. Structures should be digitally networked together so that the actions of one building and the unfavorable consequences of that action become linked, recognized, and avoided.

## References


1. Snow M, Prasad D (2011) Climate change adaptation for building designers: an introduction
2. Kuismanen K (2008) Climate-conscious architecture: design and wind testing method for climates in change. University of Oulu
3. Oxford Dictionary (2020) Climate change. https://en.oxforddictionaries.com/definition/climate_change. Accessed 22 Apr 2020
4. Callery S, Bailey D (2020) Global climate change. https://climate.nasa.gov/. Accessed 22 Apr 2020
5. NOAA (2020) Ocean acidification. https://www.noaa.gov/education/resource-collections/ocean-coasts-education-resources/ocean-acidification
6. Pachauri RK, Meyer LA (2014) Climate change 2014. Synthesis report. Contribution of working groups I, II and III to the fifth assessment report of the IPCC. IPCC Geneva, Switzerland
7. IPCC 4 (2007) Climate change 2007. The physical science basis. Cambridge University Press, Cambridge, UK, and New York, USA
8. IPCC 6 (2018) Summary for policymakers. In: Masson-Delmotte V, Zhai P, Pörtner H-O et al (eds) Global warming of 1.5 °C. An IPCC special report on the impacts of global warming of 1.5 °C above pre-industrial levels and related global greenhouse gas emission pathways, in the context of strengthening the global response to the threat of climate change. WMO Geneva, Geneva, Switzerland
9. Climate Change Institute (2022) Monthly reanalysis maps. http://cci-reanalyzer.org/reanalysis/monthly_maps/index.php. Accessed 6 Feb 2022
10. Christensen JH, Kanikicharla KK, Aldrian E et al (2013) Climate phenomena and their relevance for future regional climate change. In: Stocker TF, Qin D, Plattner G-K et al (eds) Climate change 2013: the physical science basis. Working group I contribution to the fifth

assessment report of the intergovernmental panel on climate change. Cambridge University Press, Cambridge, UK, and New York, USA, pp 1217–1308
11. SMHI (2017) New climate projections indicate more extreme weather. https://www.smhi.se/en/research/research-news/new-climate-projections-indicate-more-extreme-weather-1.12924. Accessed 22 Jul 2017
12. Eichelberger S, McCaa J, Nijssen B, Wood A (2008) Climate change effects on wind speed. North Am Wind Power 1–4
13. Nikulin G, Kjellström E, Hansson U et al (2011) Evaluation and future projections of temperature, precipitation and wind extremes over Europe in an ensemble of regional climate simulations. Tellus Ser A Dyn Meteorol Oceanogr 63:41–55. https://doi.org/10.1111/j.1600-0870.2010.00466.x
14. Forzieri G, Feyen L, Russo S et al (2016) Multi-hazard assessment in Europe under climate change. Clim Change 137:105–119. https://doi.org/10.1007/s10584-016-1661-x
15. Hansen SO (2012) Wind loading design codes. In: Fifty years of wind engineering: prestige lectures from the sixth European and African conference on wind engineering. University of Birmingham, pp 35–68
16. Stephenson DB (2008) Definition, diagnosis, and origin of extreme weather and climate events. Clim Extrem Soc 9780521870:11–23. https://doi.org/10.1017/CBO9780511535840.003
17. Cutter S, Osman-Elasha B, Campbell J et al (2012) National systems for managing the risks from climate extremes and disasters. In: Field CB, Barros V, Stocker TF et al (eds) Managing the risks of extreme events and disasters to advance climate change adaptation: special report of the intergovernmental panel on climate change. Cambridge University Press, Cambridge, UK, and New York, USA, pp 291–338
18. Sung DK (2008) Skin deep: breathing life into the layer between man and nature
19. Doumpioti C, Greenberg EL, Karatzas K (2010) Embedded intelligence: material responsiveness in façade systems. In: Life information, 20th annual conference of the association for computer-aided design in architecture ACADIA 2010, pp 258–262
20. Nagy Z, Svetozarevic B, Jayathissa P et al (2016) The adaptive solar facade: from concept to prototypes. Front Archit Res 5:143–156. https://doi.org/10.1016/j.foar.2016.03.002
21. Kuismanen K (2007) Climate and ventilation. Econo 4
22. Kerestes JF (2014) Design out of necessity: architectural approach to extreme climatic conditions. In: SIGRADI 2014: design in freedom. Blucher design proceedings, pp 130–133
23. Edwards B (2010) Rough guide to sustainability. RIBA Publishing, London, UK
24. Stathopoulos T (2009) Wind and comfort. In: Proceedings of the 5th European & African conference on wind engineering EACWE. Florence, Italy, pp 1–16
25. Goudie AS (2017) Mega-Yardangs: a global analysis. Geogr Compass 1:65–81
26. Pellitteri G, Lattuca R, Concialdi S et al (2009) Architectural shape generating through environmental forces. In: Joining languages, cultures and visions: proceedings of the 13th international CAAD futures conference. Les Presses de Université de Montréal, Montréal, Canada, pp 875–886
27. Kabošová L, Katunský D, Kmet S (2020) Wind-based parametric design in the changing climate. Appl Sci 10:1–18. https://doi.org/10.3390/app10238603
28. Jin H, Liu Z, Jin Y et al (2017) The effects of residential area building layout on outdoor wind environment at the pedestrian level in severe cold regions of China. Sustainability 9:1–18. https://doi.org/10.3390/su9122310
29. Eurocode 1 (2010) Actions on structures, part 1-4: general actions. Wind actions. Brussels
30. Tsou J-Y (1998) Applying computational fluid dynamics to architectural design development. In: Sasada T, Yamaguchi S, Morozumi M, Kaga A, Homma R (ed) Proceedings of The third conference on computer aided architectural design research in Asia. Osaka, Japan, pp 133–142
31. Cóstola D, Blocken B, Hensen JLM (2009) Overview of pressure coefficient data in building energy simulation and airflow network programs. Build Environ 44:2027–2036. https://doi.org/10.1016/j.buildenv.2009.02.006

32. Giacchetti A, Bartoli G, Mannini C (2019) Wind effects on permeable tall building envelopes: issues and potentialities. CTBUH J 20–27
33. Szücs Á (2013) Wind comfort in a public urban space—case study within Dublin Docklands. Front Archit Res 2:50–66. https://doi.org/10.1016/j.foar.2012.12.002
34. NEN (2006) Wind comfort and wind danger in the built environment: Dutch wind nuisance standard
35. Du Y, Mak CM, Kwok K et al (2017) New criteria for assessing low wind environment at pedestrian level in Hong Kong. Build Environ 123:23–36. https://doi.org/10.1016/j.buildenv.2017.06.036
36. Pirner M, Fischer O (2003) Zatížení staveb větrem. Praha
37. Baghaei Daemei A, Khotbehsara EM, Nobarani EM, Bahrami P (2019) Study on wind aerodynamic and flow characteristics of triangular-shaped tall buildings and CFD simulation in order to assess drag coefficient. Ain Shams Eng J 10:541–548. https://doi.org/10.1016/j.asej.2018.08.008
38. Glenn Research Center (2020) Shape effects on drag. https://www.grc.nasa.gov/WWW/K-12/airplane/shaped.html. Accessed 9 May 2020
39. Kaushik V, Janssen P (2015) Urban wind flow: investigating the use of animation software for simulating windflow around buildings. In: Real time—proceedings of the 33rd eCAADe conference, vol 1, pp 225–234
40. Grobman YJ, Yezioro A, Capeluto IG (2009) Computer-based form generation in architectural design—a critical review. Int J Archit Comput. https://doi.org/10.1260/1478-0771.7.4.535
41. Veenendaal D, Block P (2012) An overview and comparison of structural form finding methods for general networks. Int J Solids Struct 49:3741–3753. https://doi.org/10.1016/j.ijsolstr.2012.08.008
42. Hu Y, Li Q (2014) Study on an architect-oriented approach of structural performance-based design. In: SIGgraDi 2014: design in freedom. Blucher design proceedings, Montevideo, Uruguay, pp 111–114
43. Kuzmin D (2010) A guide to numerical methods for transport equations. Friedrich-Alexander-Universität, Erlangen-Nürnberg
44. Moukalled F, Mangani L, Darwish M (2016) Fluid mechanics and its applications. Springer
45. Malkawi AM, Srinivasan RS, Yi YK, Choudhary R (2005) Decision support and design evolution: integrating genetic algorithms, CFD and visualization. Autom Constr 14:33–44. https://doi.org/10.1016/j.autcon.2004.06.004
46. Mazauric L, Demers CM, Potvin A (2018) Climate form finding for architectural inhabitability. Ambiances 4:1–29. https://doi.org/10.4000/ambiances.1688
47. Taleb H, Musleh MA (2015) Applying urban parametric design optimisation processes to a hot climate: case study of the UAE. Sustain Cities Soc 14:236–253. https://doi.org/10.1016/j.scs.2014.09.001
48. Erhan H, Wang I, Shireen N (2014) Interacting with thousands: a parametric-space exploration method in generative design. In: ACADIA 2014: design agency. Los Angeles, California, pp 619–625
49. Bassolino E, Ambrosini L (2016) Parametric environmental climate adaptive design: the role of data design to control urban regeneration project of Borgo Antignano, Naples. Procedia Soc Behav Sci 216:948–959. https://doi.org/10.1016/j.sbspro.2015.12.092
50. Barlieb C, Richter C, Greschner B, Tamke M (2009) Whispering wind: digital practice and the sustainable agenda. In: Proceedings of the 27th eCAADe conference. Istanbul, Turkey, pp 543–550
51. Hartog JP den, Koutamanis A, Luscuere PG (2000) Possibilities and limitations of CFD simulation for indoor climate analysis. In: Fifth design and decision support systems in architecture and urban planning—part one: architecture proceedings. Nijkerk, Netherlands, pp 152–167
52. Wang L, Tang Z, Ji G (2016) Toward the wind-related building performative design. In: Chien S, Choo S, Schnabel MA et al (eds) Living systems and micro-Utopias: towards continuous designing proceedings of the 21st international conference of the association for computer-aided architectural design research in Asia CAADRIA. Hong Kong, pp 209–218

53. Kuenstle MW (2001) Computational flow dynamic applications in wind engineering for the design of building structures in wind hazard prone urban areas. In: Proceedings of the 5th SIGradi conference. Concepcion, Chile, pp 67–70
54. Kormaníková L, Achten H, Kopřiva M, Kmeť S (2018) Parametric wind design. Front Archit Res 7:383–394. https://doi.org/10.1016/j.foar.2018.06.005
55. Chung DHJ, Choo M-LL (2010) Computational fluid dynamics for urban design: the prospects for greater integration. Int J Archit Comput 09:33–53
56. Shaeri J, Aflaki A, Yaghoubi M, Janalizadeh H (2018) Investigation of passive design strategies in a traditional urban neighborhood: a case study. Urban Clim 26:31–50. https://doi.org/10.1016/j.uclim.2018.08.003
57. Kim HJ, Kim JS (2018) Design methodology for street-oriented block housing considering daylight and natural ventilation. Sustain 10:1–22. https://doi.org/10.3390/su10093154
58. Chronis A, Liapi KA, Sibetheros I (2012) A parametric approach to the bioclimatic design of large scale projects: the case of a student housing complex. Autom Constr 22:24–35. https://doi.org/10.1016/j.autcon.2011.09.007
59. Stavridou AD (2015) Breathing architecture: conceptual architectural design based on the investigation into the natural ventilation of buildings. Front Archit Res 4:127–145. https://doi.org/10.1016/j.foar.2015.03.001
60. Yuan C, Shan R, Zhang Y et al (2019) Multilayer urban canopy modelling and mapping for traffic pollutant dispersion at high density urban areas. Sci Total Environ 647:255–267. https://doi.org/10.1016/j.scitotenv.2018.07.409
61. Zhang A, Xia C, Li W (2022) Exploring the effects of 3D urban form on urban air quality: evidence from fifteen megacities in China. Sustain Cities Soc 78:1–18. https://doi.org/10.1016/j.scs.2021.103649
62. Xu A, Zhao RH (2020) Wind-resistant structural optimization of a supertall building with complex structural system. Struct Multidiscip Optim 62:3493–3506. https://doi.org/10.1007/s00158-020-02652-x
63. Asghari Mooneghi M, Kargarmoakhar R (2016) Aerodynamic mitigation and shape optimization of buildings: review. J Build Eng 6:225–235. https://doi.org/10.1016/j.jobe.2016.01.009
64. Kabošová L, Foged I, Kmeť S, Katunský D (2019) Hybrid design method for wind-adaptive architecture. Int J Archit Comput 17:1–16. https://doi.org/10.1177/1478077119886528
65. Kuenstle MW (2002) Flow structure environment simulation: a comparative analysis of wind flow phenomena and building structure interaction. In: Koszewski K, Wrona S (ed) 20th eCAADe conference proceedings. Warsaw, pp 564–568
66. Kabošová L, Chronis A, Galanos T, Kmeť S, Katunský D (2022) Shape optimization during design for improving outdoor wind comfort and solar radiation in cities. Build Environ, 226:109668. https://doi.org/10.1016/j.buildenv.2022.109668
67. Poćwierz M, Zielonko-Jung K (2021) An analysis of wind conditions at pedestrian level in the selected types of multi-family housing developments. Environ Fluid Mech 21:83–101. https://doi.org/10.1007/s10652-020-09763-5
68. Athanailidi P, Fatah Gen Schieck A, Tenu V, Chronis A (2014) Tensegrity systems acting as windbreaks: form finding and fast fluid dynamics analysis to address wind funnel effect. In: Proceedings of SimAUD 2014. Tampa, Florida, USA, pp 9–16
69. Moya Castro RA (2015) Wind analysis in the early design stage: an empirical study of wind visualisation techniques for architects. RMIT University
70. Kalantar N, Borhani A (2017) Breathable walls—computational thinking in early design education. In: Protocols, flows, and glitches—proceedings of the 22nd CAADRIA conference. Xi'an Jiaotong-Liverpool University, Suzhou, China, pp 377–386
71. Ishugah TF, Li Y, Wang RZ, Kiplagat JK (2014) Advances in wind energy resource exploitation in urban environment: a review. Renew Sustain Energy Rev 37:613–626. https://doi.org/10.1016/j.rser.2014.05.053
72. Zhou H, Lu Y, Liu X et al (2017) Harvesting wind energy in low-rise residential buildings: Design and optimization of building forms. J Clean Prod 167:306–316. https://doi.org/10.1016/j.jclepro.2017.08.166

73. Vatansever D, Siores E, Shah T (2012) Alternative resources for renewable energy: piezoelectric and photovoltaic smart structures. Glob Warm Impacts Futur Perspect. https://doi.org/10.5772/50570
74. Cimmino MC, Miranda R, Sicignano E et al (2017) Composite solar façades and wind generators with tensegrity architecture. Compos Part B Eng 115:275–281. https://doi.org/10.1016/j.compositesb.2016.09.077
75. Janssen W, Blocken B, Van Hooff T (2013) Use of CFD simulations to improve the pedestrian wind comfort around a high-rise building in a complex urban area. In: 13th Conference of international building performance simulation association, Chambéry, France, 26–28 August. Chambéry, France, pp 1918–1925
76. Zahid Iqbal QM, Chan ALS (2016) Pedestrian level wind environment assessment around group of high-rise cross-shaped buildings: effect of building shape, separation and orientation. Build Environ 101:45–63. https://doi.org/10.1016/j.buildenv.2016.02.015
77. Abu-Zidan Y, Mendis P, Gunawardena T (2021) Optimising the computational domain size in CFD simulations of tall buildings. Heliyon, vol 4, 7:e06723. https://doi.org/10.1016/j.heliyon.2021.e06723
78. Shen Q, Uselton S, Pang A (2003) Comparisons of wind tunnel experiments and computational fluid dynamics simulations. J Vis 6:31–39. https://doi.org/10.1007/BF03180962
79. Kieferle J, Wössner U, Becker M (2007) Interactive simulation in virtual environments—a design tool for planners and architects. Int J Archit Comput 5:115–126. https://doi.org/10.1260/147807707780912903
80. Kaijima S, Bouffanais R, Willcox K (2013) Computational fluid dynamics for architectural design. In: Stouffs R, Janssen PHT, Roudavski S, Tunçer B (eds) Proceedings of the 18th international conference of the association of computer-aided architectural design research in Asia CAADRIA 2013. Hong Kong
81. Chronis A, Dubor A, Cabay E, Roudsari MS (2017) Integration of CFD in computational design: an evaluation of the current state of the art. In: Fioravanti A, Cursi S, Elahmar S et al (eds) Proceedings of the 35th international conference eCAADe: ShoCK! Sharing of computable knowledge! Sapienza University of Rome, Rome, Italy, pp 601–610
82. Sanquer S, Caniot G, Bandhare S (2015) Wind assessment in urban area with cfd tools application to natural ventilation potential and outdoor pedestrian comfort. In: Proceedings of BS2015: 14th conference of international building performance simulation, pp 1684–1691
83. Maffessanti V (2020) Wind and urban spaces. Evaluation of a CFD parametric framework for early stage design. Build Simul Appl 2020-June:67–76
84. Carmichael A, Ahmed S, Dyer R et al (2020) RhinoCFD user guide
85. ODS Engineering (2020) Swift: building physics for grasshopper. https://www.ods-engineering.com/tools/ods-swift/. Accessed 4 May 2020
86. Chronis A, Turner A, Tsigkari M (2011) Generative fluid dynamics: integration of fast fluid dynamics and genetic algorithms for wind loading optimization of free form surfaces. SimAUD2011 79–85
87. Waibel C, Bystricky L, Kubilay A et al (2017) Validation of grasshopper-based fast fluid dynamics for air flow around buildings in early design stage. 15th IBPSA Conf 2147–2156. https://doi.org/10.26868/25222708.2017.582
88. Galanos T, Chronis A (2022) Time for change—the InFraRed revolution: how AI-driven tools can reinvent design for everyone. Archit Des 92:108–115. https://doi.org/10.1002/ad.2821
89. Fink T, Koenig R, Duering S, Chronis A (2022) Environmental performance assessment-the optimisation of high-rises in Vienna environmental performance assessment the optimisation of high-rises in Vienna
90. Kabošová L, Chronis A, Galanos T (2021) Leveraging urban configurations for achieving wind comfort in cities. In: Gomez P, Braida F (eds) SIGgraDi 2021: designing possibilities. Blucher design proceedings, Brazil/Georgia/online, pp 79–90
91. Waibel C, Zhang R, Wortmann T (2021) Physics meets machine learning: coupling FFD with regression models for wind pressure prediction on high-rise facades. Association for Computing Machinery

92. Alexander DK, Jenkins HG, Jones PJ (1997) A comparison of wind tunnel and CFD methods applied to natural ventilation design. In: Building simulation: proceedings of the IBPSA conference. Prague, Czech Republic, pp 1–7
93. Olekšáková I, Hubová O, Konečná L (2015) Influence of nearby structure on the wind flow around the cube structure. Rocz Inz Bud 15:29–36
94. Moya Castro RA, Prohasky D, Watkins S et al (2014) Aerodynamic strategy applied in an urban shelter design—simulation and analysis of aerodynamic phenomena in an urban context. In: eCAADe 2014 fusion—data integration its best, vol 1, pp 137–144
95. Sterk T d'Estree (2003) Using actuated tensegrity structures to produce a responsive architecture. In: ACADIA22: connecting crossroads of digital discourse. Ball State University, Indianapolis, Indiana, pp 84–93
96. Biloria N, Sumini V (2009) Performative building skin systems: a morphogenomic approach towards developing real-time adaptive building skin systems. Int J Archit Comput 7:643–675. https://doi.org/10.1260/1478-0771.7.4.643
97. Persiani S, Battisti A, Persiani SGL, Wolf T (2016) Autoreactive architectural facades-discussing unpowered kinetic building skins and the method of evolutionary optimization. In: Proceedings of 11th conference on adaptive building skins
98. Orhon AV (2016) Adaptive building shells. In: Efe R, Matchavariani L, Yaldır A, Lévai L (eds) Developments in science and engineering. St. Kliment Ohridski University Press, pp 555–567
99. Badarnah L (2017) Form follows environment: biomimetic approaches to building envelope design for environmental adaptation. Buildings 7. https://doi.org/10.3390/buildings7020040
100. Senatore G, Duffour P, Winslow P (2018) Energy and cost assessment of adaptive structures: case studies. J Struct Eng (United States) 144:1–23. https://doi.org/10.1061/(ASCE)ST.1943-541X.0002075
101. Abdelmohsen S, Massoud P, Elshafei A (2016) Using tensegrity and folding to generate soft responsive architectural skins. In: Complexity and simplicity: proceedings of the 34th international conference on education and research in computer aided architectural design in Europe. University of Oulu, Oulu, Finland, pp 529–536
102. Körner A, Mader A, Saffarian S, Knippers J (2016) Bio-inspired kinetic curved-line folding for architectural applications. ACADIA 2016 posthuman front data. Des Cogn Mach—Proceedings of 36th annual conference of the association for computer-aided design in architecture, pp 270–279
103. Holstov A, Bridgens B, Farmer G (2015) Hygromorphic materials for sustainable responsive architecture. Constr Build Mater 98:570–582. https://doi.org/10.1016/j.conbuildmat.2015.08.136
104. Reichert S, Menges A, Correa D (2015) Meteorosensitive architecture: biomimetic building skins based on materially embedded and hygroscopically enabled responsiveness. Comput Des 60:50–69. https://doi.org/10.1016/j.cad.2014.02.010
105. Foged IW, Pasold A (2015) thermal activated envelope: a method and model for embedding behaviour in a responsive envelope by bi-materials. In: Real time extending the reach of computation: proceedings of the 33rd eCAADe conference. Vienna, pp 449–458
106. Schleicher S, Lienhard J, Poppinga S et al (2015) A methodology for transferring principles of plant movements to elastic systems in architecture. CAD Comput Aided Des 60:105–117. https://doi.org/10.1016/j.cad.2014.01.005
107. Körner A, Born L, Mader A et al (2018) Flectofold—a biomimetic compliant shading device for complex free form facades. Smart Mater Struct 27. https://doi.org/10.1088/1361-665X/aa9c2f
108. Nicholas P, Tamke M (2013) Computational strategies for the architectural design of bending active structures. Int J Sp Struct 28:215–228. https://doi.org/10.1260/0266-3511.28.3-4.215
109. Deleuran AH, Schmeck M, Quinn G et al (2015) The tower: modelling, analysis and construction of bending active tensile membrane hybrid structures. In: Proceedings of international association for shell and spatial structures

110. Kontovourkis O, Phocas MC, Alexandrou KC, Frangogiannopoulos S (2017) Configuration and deformation control of a hybrid cable bending-active structure. Int J Comput Methods Exp Meas 5:475–483. https://doi.org/10.2495/CMEM-V5-N4-475-483
111. Motro R (2003) Tensegrity: structural systems for the future, 1st edn. Butterworth-Heinemann
112. Khoo CK, Salim F, Burry J (2011) Designing architectural morphing skins with elastic modular systems. Int J Archit Comput 9:397–419. https://doi.org/10.1260/1478-0771.9.4.397
113. Sushant V, Pradeep D (2013) Adaptive [skins]: responsive building skin systems based on tensegrity principles. In: Future traditions [1st eCAADe regional international workshop proceedings]. FAUP Publicações, Porto, Portugal, pp 155–170
114. Motro R (2011) Tension structures. In: Fifty years of progress for shell and spatial structures. Multi-Science Publishing, pp 1–14
115. Peña DM, Llorens I, Sastre R (2010) Application of the tensegrity principles on tensile textile constructions. Int J Sp Struct 25:57–67. https://doi.org/10.1260/0266-3511.25.1.57
116. Cretu SM (2011) Innovative design in tensegrity field. Procedia Eng 9:261–269. https://doi.org/10.1016/j.proeng.2011.03.117
117. Kmet S, Platko P, Mojdis M (2012) Analysis of adaptive light-weight structures. Procedia Eng 40:199–204. https://doi.org/10.1016/j.proeng.2012.07.080
118. Spisak M, Kmet S (2017) Shape and stress modification of a chosen tensegrity system. Procedia Eng 190:637–644. https://doi.org/10.1016/j.proeng.2017.05.391
119 Chan WL, Arbelaez D, Bossens F, Skelton RE (2004) Active vibration control of a three-stage tensegrity structure. In: Proceedings of SPIE 5386, smart structures and materials 2004: damping and isolation. San Diego, CA, pp 1–2
120. Yang S, Sultan C (2016) Modeling of tensegrity-membrane systems. Int J Solids Struct 82:125–143. https://doi.org/10.1016/j.ijsolstr.2015.09.012
121. Jun JW, Silverio M, Llubia JA et al (2017) Remembrane. In: CAAD futures 17. Istanbul, Turkey, pp 180–198
122. Phocas MC, Kontovourkis O, Nicolaou N (2014) Design concept of a kinetic form-active hybrid envelope structure. Int J Des Nat Ecodynamics 9:13–30. https://doi.org/10.2495/DNE-V9-N1-13-300
123. Bertin TB (2013) Evaluating the use of particle-spring systems in the conceptual design of grid shell structures. MIT

# Chapter 2
# Wind-Driven Design

This chapter contains two sub-chapters:

1. Description of the method
2. Wind-driven design in Stockholm's docks

## 2.1 Description of the Method

The proposed wind-driven method employs Research Through Design (RTD) methodological approach, which actively engages designing in the research process. The benefits of the RTD include examining proposed designs, predicting their effects, and optimizing them before construction, in the early design stage. Within the context of the cities' adaptation to climate change, the iterative character of RTD incorporated into the design practice, coupled with CFD for anticipating the wind flow, is a more fitting approach than employing the expensive and time-consuming wind tunnel tests [1]. Simulations engaged in the early design stage provide a 'copy' of reality which helps the designer understand climate information (such as the influence of the wind conditions on the built environment) and, based on the gained knowledge make optimal design decisions [2]. In the first part of this book, wind simulations will be utilized during early-stage architectural design. In the next chapter, 3D prototyping will be combined with digital simulations to create a wind-adaptive envelope and investigate its behavior in the wind.

### *2.1.1 Steps of the Method*

Despite the impact of urban context and architecture on wind fluxes, which was mentioned before, the wind is still not one of the crucial factors in architectural design. The book focuses on the wind as a form-determining element in the architectural

 
L. Kabošová et al., *Designing with the Wind*, Digital Innovations in Architecture, Engineering and Construction, https://doi.org/10.1007/978-3-031-24441-4_2

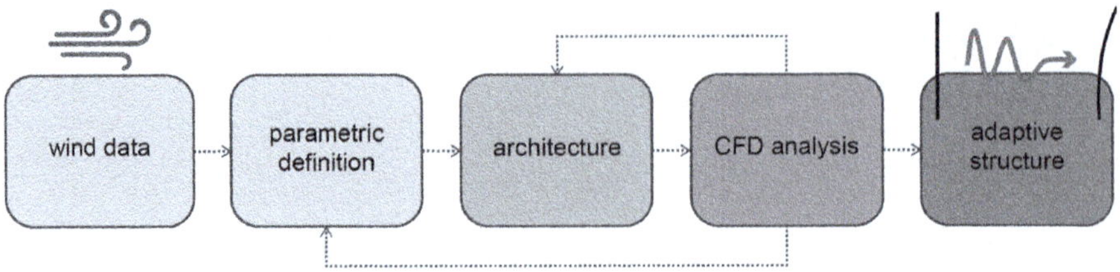


**Fig. 2.1** The steps of the design method and their relations

shape-generating process. The design method is a recursive loop, leading toward the wind-derived architecture (Fig. 2.1).

#### 2.1.1.1 Wind Data: Analysis of the Local Wind Microclimate

The wind data are derived from multiple sources, primarily from the EnergyPlus web database and Lobelia Earth, Barcelona's recent research project [3], both available online. The digital EnergyPlus data, well-fitted for energy simulations, are obtained as an *epw weather file and imported into Grasshopper. The data set comprises archived recordings from meteorological stations going back 25 years [4]. Employing the Ladybug[1] add-on (Fig. 2.2), the weather file (*epw) can be converted into the graphical interpretation through wind-rose, with the annual and monthly wind speeds, directions, as well as frequency. The Lobelia Earth's data source is the ERA5 dataset, with the data from 1981 to 2010.

#### 2.1.1.2 Parametric Definition: Designing with Grasshopper

Nature-driven architectural solutions can be intricate, but the parametric design, coupled with advanced simulation tools, yields a comprehension of the mutual relations between the planned architecture and its surroundings before the final design stage and the realization. The unfavorable effects of the wind manifest differently in various climates. Cold climates can be exposed to cold winds with extreme wind speeds, resulting in extreme wind loads on buildings, usually leading to robust building structures. The wind in arid climates can cause wind damage to buildings, including sand and dust particles that harm buildings in strong winds. The wind speed and temperature influence pedestrian comfort due to the cold, accelerated wind, or on the contrary, no wind in dry and hot climates.

As an introduction to the wind-driven design method, this book establishes basic categories based on the reciprocal interactions of the wind and the urban environment

[1] https://www.ladybug.tools/index.html.

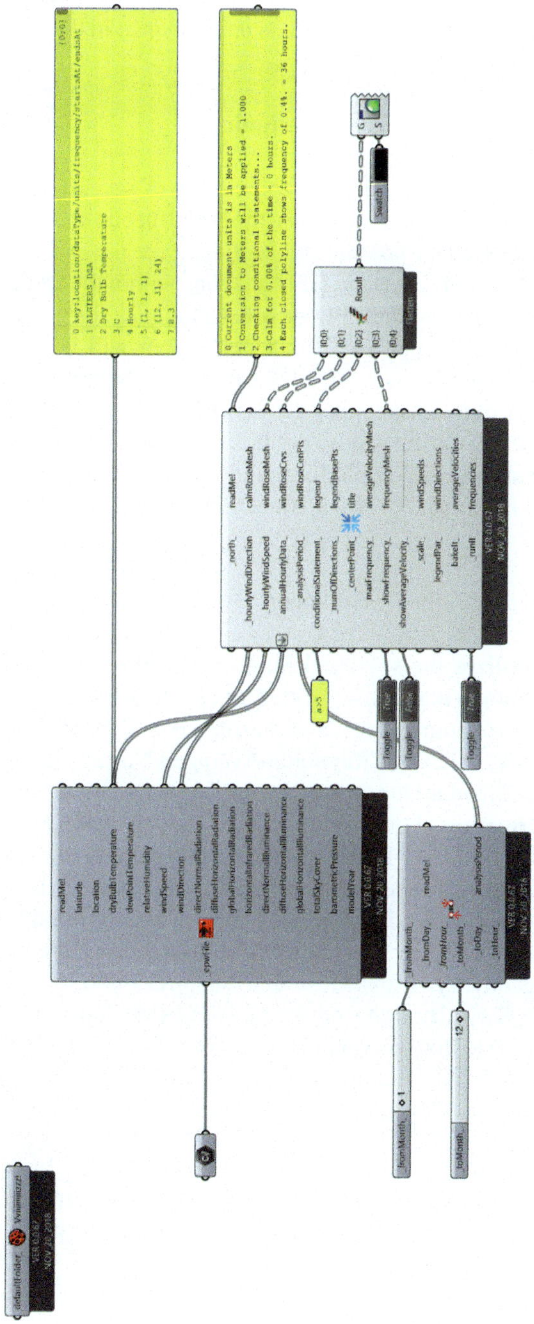


**Fig. 2.2** Grasshopper script employing Ladybug for generating the wind-rose

(Fig. 2.3). Based on the specified design goal, the parametric definition is subsequentially developed, with individual parametric variables, according to the design intentions, which affect the final design and its performance in the wind (Table 2.1).

#### 2.1.1.3 CFD Analysis: Wind-Driven Designing

The first design stages are usually about explorations of several design directions, so investigating multiple design variants in the wind needs to be as effective as possible. As was determined earlier, the most efficient for the early design stage, in terms of cost, time, and accuracy, is the CFD analysis within the same 3D modeling environment, complemented by the FFD or ML techniques for wind prediction. In the next chapter, Rhino CFD and Swift for Grasshopper will be employed within the application study. The approach enables parametrically setting the VWT and boundary conditions in one Grasshopper canvas (Fig. 2.4). It works like Butterfly[2] for Grasshopper, however, for this study, Swift is selected. Both Swift and Butterfly are based on the OpenFOAM CFD platform and require the installation of either a Linux-Ubuntu virtual machine (Swift) or BlueCFD Core[3] (Butterfly).

Swift (Procedural Compute) for Grasshopper

Swift CFD analysis utilizes the so-called SimpleFoam (Semi-Implicit Method for Pressure-Linked Equations) iterative algorithm and RAS (Reynolds-Averaged Simulation) turbulent model for computing Navier–Stokes equations [5]. RAS, used in the OpenFOAM framework, is RANS (Reynolds-Averaged Navier–Stokes) model. The simulations exploit the Standard k-ε turbulence model for incompressible steady-state flows. For more information about the turbulence models and CFD theory, please refer to [6–9]. In the OpenFOAM-based simulation, the number of iterations, convergence criteria, or both are set to determine when the calculations stop. Convergence criteria for calculated initial residuals of pressure $p$, velocity in all its components $v_{x,y,z}$, the turbulent kinetic energy $k$, and turbulent dissipation $\varepsilon$ are determined. For meshing the tested geometry, BlockMesh and SnappyHexMesh are employed. A crucial remark is that Swift, running on the OpenFOAM platform, uses kinematic pressure in incompressible flow solvers [5]:

$$p_{\mathrm{k}} = \frac{p_{\mathrm{a}}}{\rho} \tag{2.1}$$

where $p_{\mathrm{k}}$ is the kinematic pressure ($m^2/s^2$), $p_{\mathrm{a}}$ is the static pressure (Pa), and $\rho$ is the density of the flow ($kg/m^3$). Hence, the units $m^2/s^2$ will be primarily used and will

[2] https://www.ladybug.tools/butterfly.html.

[3] https://github.com/ladybug-tools/butterfly/wiki/1.-blueCFD-Core-(OpenFOAM)-Installation.

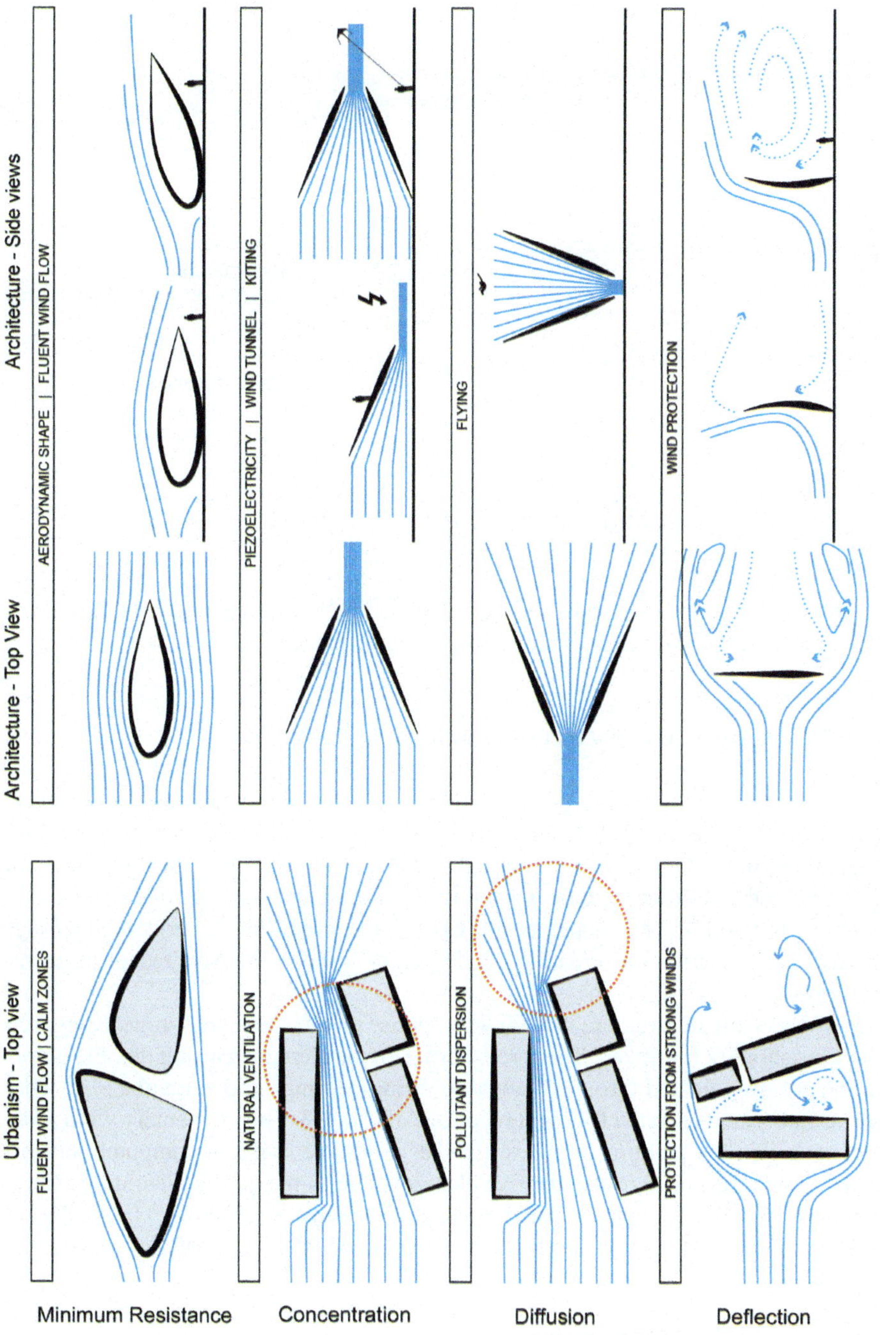


**Fig. 2.3** The effect of the urban morphology and architectural shape on the wind flow

**Table 2.1** The goals of wind-driven designing and possible variables in parametric designing

| | Favorable wind effects achieved through architecture | Parametric variables |
|---|---|---|
| Minimum resistance | • Wind pressure reduction with aerodynamic shape designs with low values of drag coefficient $c_d$<br>• Lighter building structures<br>• Undisturbed wind flow | • Height h and length l of the buildings<br>• The frontal area on the windward side (first wind impact) |
| Concentration | • Natural ventilation<br>• Wind energy harvesting | • Specific dimensions of openings (also in urban scale, spaces between buildings)<br>• The relative position of the openings |
| Diffusion | • Pollutant dispersion | • Specific dimensions of openings (also in urban scale, spaces between buildings)<br>• The relative position of the openings |
| Deflection | • Shelters from the wind<br>• Pedestrian wind comfort | • Height h<br>• Angle $\alpha$ |

be subsequently converted to Pa, multiplying the kinematic pressure by 1.25 kg/m$^3$, which represents the density of air, used in the Eurocode 1.

#### 2.1.1.4 Adaptive Architecture: Responding to the Wind

The proposed wind-driven design method fuses parametric designing with parametrically-controlled CFD simulations to create a design loop, enabling building shape or urban layout optimization based on the wind. Once the most suitable urban layout and the best-fitting architectural form are found, the next step of the method is designing the building envelope, morphing with the wind flow. A networked tensegrity and tensile membrane, creating a lightweight, wind-responsive façade element, is proposed.

For observing the structure's dynamic response to the wind in real-time, Kangaroo 2 for Grasshopper is utilized. The forces acting on the element causing flexible deformation are investigated through Kangaroo 2 Engineering[4] and Karamba 3D.[5] That will be elaborated on in the following sections of Chap. 3. The momentary wind pressure, acting on the building surfaces, induces a passive (without computer control) and reversible response of its adaptive elements due to the geometric configuration of tensegrity and the material characteristics of the tensile membrane. In the digital analysis of the real-time shape-change in the wind, the value of the pressure *p*,

[4] https://github.com/CecilieBrandt/K2Engineering.

[5] https://www.karamba3d.com/.

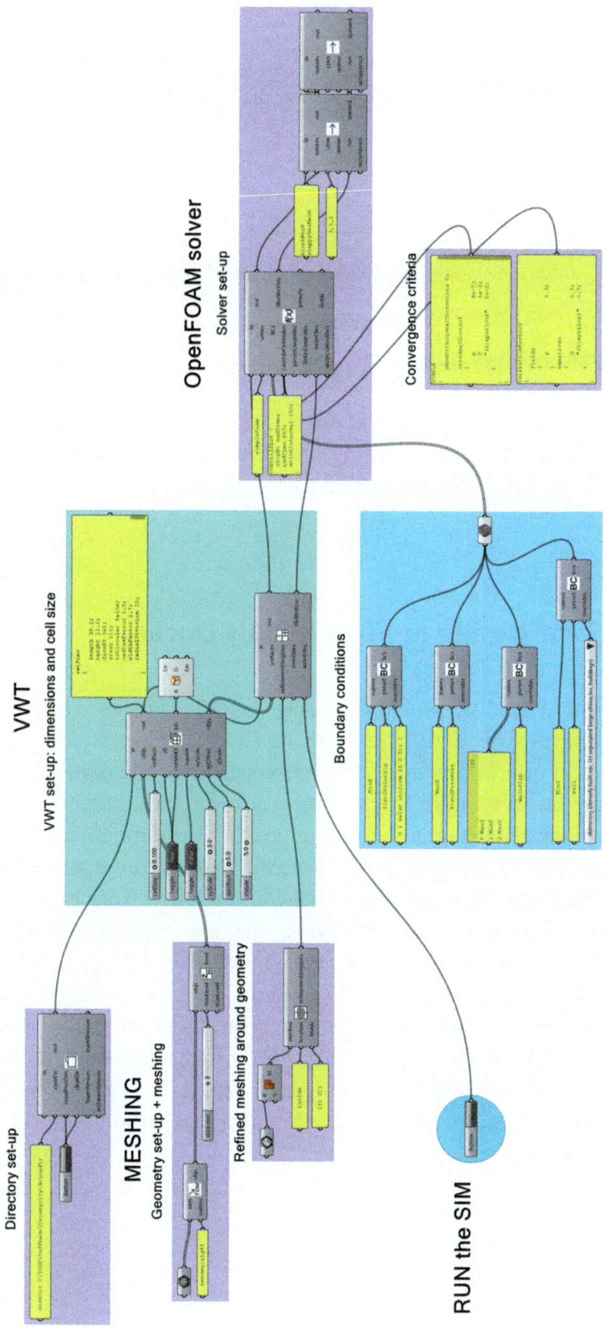


**Fig. 2.4** The CFD simulation set-up in Swift for Grasshopper

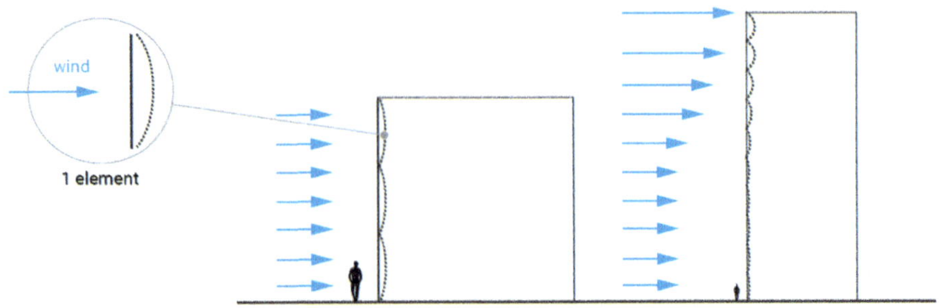


**Fig. 2.5** The scheme of the wind-adaptive behavior of intended adaptive building skin

representing the wind acting on the structure, is derived from the wind speed using Bernoulli's equation (Eq. 1.2).

The tensegrity-membrane envelope applied instead of a building roof or façade creates a dynamic, shape-changing building skin, morphing according to the intensity of the acting wind (Fig. 2.5). With the change in the shape, the wind pattern and wind pressure are altered.

## 2.2 Wind-Driven Design in Stockholm's Docks

Situated on the eastern side of Stockholm, the Loudden Docks used to be one of the most strategic ports of the Baltic Sea, storing heating oil, petroleum, and diesel. In 2011, it stopped all operations due to the perspective of the district architectural revitalization within the Stockholm Royal Project. Except for Loudden port, a sustainable revitalization of the Hjorthagen, Värtahamnen, and Frihamnen, is planned within Stockholm's project [10]. Loudden Docks' transformation into apartment blocks is planned in the later phases. Nowadays, the brownfield, consisting of more than 100 cylindrically shaped reinforced concrete silos, spreads over an area of 10.25 ha, visually blocking access to the waterfront (Fig. 2.6).

The architectural competition from 2017 to find alternative architectural ideas for the district triggered the creation of the following application study. It approaches the site creatively, keeping, to some extent, the unique character of the site and transforming it into a post-industrial wind park (Fig. 2.7). As will be illustrated, light architectural interventions can change the turbulent wind situation caused by the clustered urban distribution of silos.

**Fig. 2.6** The Docks with cylindrical silos (*Source* https://www.flickr.com/photos/norradjurgardsstaden/32813055227/in/photostream/)

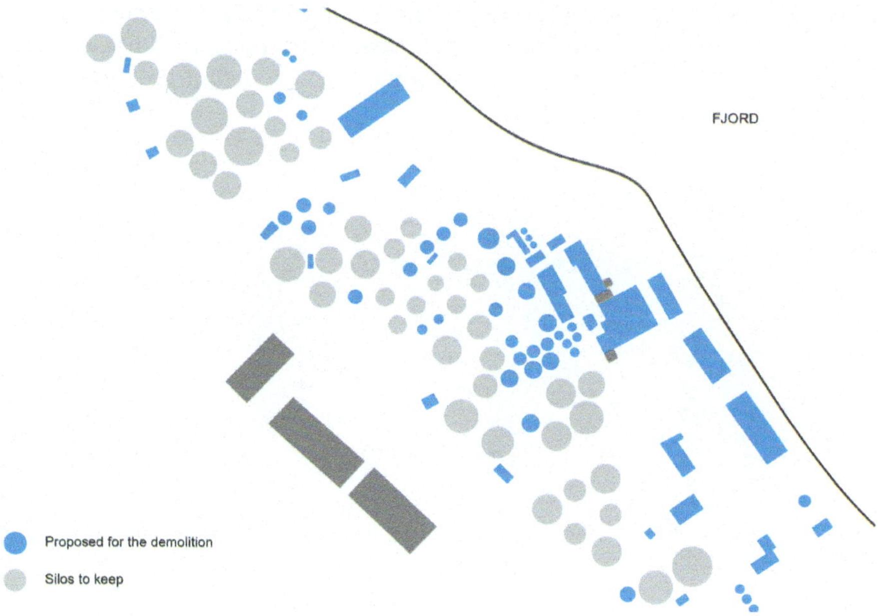


**Fig. 2.7** The layout of the Loudden Docks brownfield

### 2.2.1 Wind Data

The weather data from the meteorological station in Stockholm Arlanda airport in an *epw file format represent the weather in Loudden Docks in this study. The wind-rose generated through Ladybug in Grasshopper indicates that the westerly winds are prevailing with the annual average wind velocity of 3.4 m/s, while the gusts are southerly and northerly (Fig. 2.8). The wind speed is 18.77% of the year higher than 5 m/s, which is the set comfort limit for pedestrians according to the Dutch standard [11]. If >10% of the time this comfort limit is exceeded, this results in poor conditions for strolling and sitting activities. The Lawson criteria specify a higher threshold level; wind speeds from 5.3 m/s represent the onset of discomfort and disturb sitting or a slow walk [12, 13]. The speed of 15 m/s is exceeded 0.25% of the year (Table 2.2). Crossing this wind speed limit >0.05% of the year indicates a limited risk of wind danger for pedestrians [11]. Accompanied by cooler air temperatures between 5 and 18 °C on average, this further disturbs the pedestrian wind comfort in Stockholm.

The southerly winds are 0.55% per year stronger than 9 m/s (Fig. 2.9). Exterior spaces with the lack of sun and lower air temperatures, co-occurring with higher wind speeds, are usually avoided by people.

The weather information retrieved from the Lobelia Earth database specifies the yearly average wind speed of 4.17 m/s (Fig. 2.10). The monthly maximum wind gusts reach 58 km/h (16.1 m/s) in August and 68 km/h (18.9 m/s) in January.

The wind information is leveraged in subsequent steps of the wind-driven design. The main design goals will be (A) ensuring the wind comfort of pedestrians using

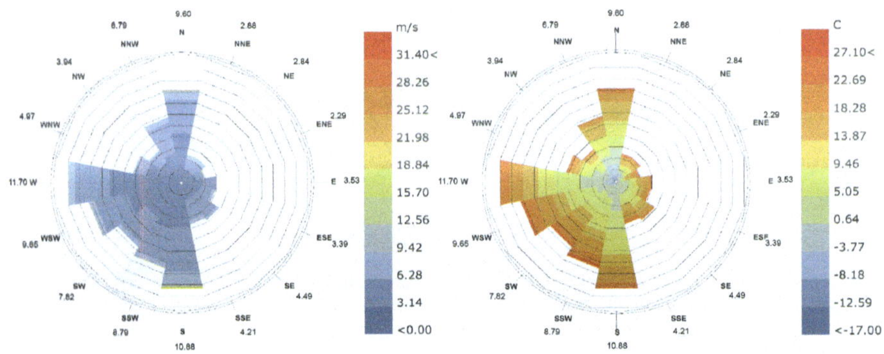


**Fig. 2.8** Wind-rose of Stockholm, Sweden, showing the wind directions and frequencies in % (*Data source* EnergyPlus)

**Table 2.2** Wind data for Stockholm's Arlanda airport

| Total wind | = 0 m/s | >5 m/s | >15 m/s | Average wind velocity |
|---|---|---|---|---|
| 8760 h (100%) | 63 h (2.24%) | 1644 h (18.77%) | 21.9 h (0.25%) | 3.4 m/s |

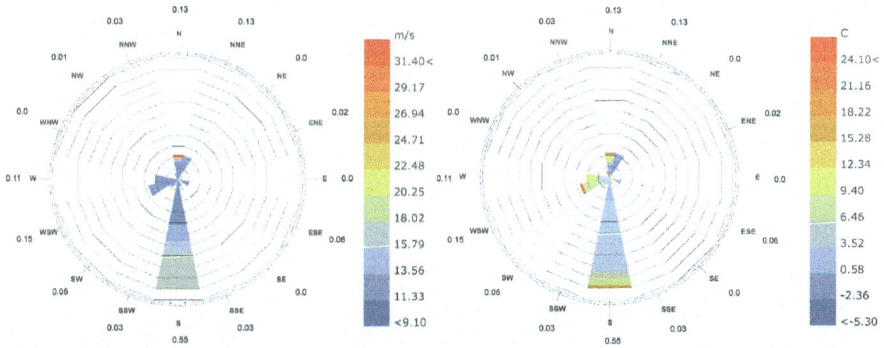


**Fig. 2.9** Wind-rose, depicting the wind speeds >9 m/s for all wind directions (*Data source* EnergyPlus)

architecture, (B) harvesting the wind energy, and (C) creating a lightweight building envelope. Therefore, three wind-speed values will be investigated:

- 5 m/s, representing the threshold value for wind comfort of pedestrians,
- 9 m/s, representing the southerly gusts,
- 24 m/s (extreme wind speed), representing the mean wind velocity $v_m$, for designing structures around Stockholm, Sweden.

### 2.2.2 *Urbanism: CFD Analysis*

The urban morphology of the application study site is unusual, as most of the existing 'buildings' are cylinders with a height and diameter between 10 and 35 m. The cylindrical shapes, along with the dense placement of the silos, create an extreme, turbulent wind situation. A forest surrounds the southwest side of the Loudden Docks brownfield. The fjord lies in the northeast. The influence of these factors on the wind flow is considered in the CFD simulations as the terrain roughness. Three old industrial buildings on the southwest affect the southerly and the westerly wind flow, so they are considered in the CFD analyses. The wind, passing through the site, is squeezed in-between the groups of silos, resulting in the flow acceleration or deflection in the horizontal and the vertical direction. The turbulent flow forming on the leeward side is a consequence. Furthermore, except for the southern part of the site, the rest of the silos create self-shadows throughout the year.

This unique and, at the same time, extreme microclimate determines the application study site and, therefore, is a driving force in the subsequent design steps.

The wind flow through the site in the existing, extreme situation is examined using the Rhino CFD tool. As the wind data suggest, the westerly winds are prevailing, and the wind gusts are typically southerly. For the CFD calculations, RANS, Chen-Kin $k$-$\varepsilon$ turbulence model is selected. First, the wind tunnel around the tested geometry is

**Fig. 2.10** The distribution of average wind speed. The colors depict the speed 10 km/h (black)—27 km/h (cyan) (*Data source* Lobelia)

created with regard to the best guidelines for the correct wind tunnel settings [14, 15]. The domain size equals 1175 × 830 × 175 m for the westerly winds and 863 × 1137 × 175 m for the southerly winds. Both simulation cases consist of approximately 2 million cells, denser around the tested geometry, bringing more detail to the areas of interest. The boundary conditions are determined through wind attribute settings. The logarithmic function is employed to vertically distribute the wind velocity, changing with the height from the ground (wind profile) [16]. The calculations are set to converge when all error values are under 1%. The simulation of the westerly winds is performed in the threshold wind speed 5 m/s, set at the reference height of 10 m above the ground. The results are evaluated in the horizontal plane, 1.75 m above the ground, to analyze the flow at the pedestrian level (Fig. 2.11).

Three buildings on the southwest act as a blockage to the wind, creating a turbulent flow that subsequently collides with the silos. The zones with the accelerated wind (A) can be recognized (green to red color), together with turbulent zones (T) and calmer zones (dark blue color). Although the silo groups can decrease the wind speed, creating calmer zones, a more detailed analysis will unveil turbulence on the leeward side of the silos.

Throughout the year, the wind gusts, stronger than 9 m/s, are almost half the time southerly. Accordingly, the simulation of the southerly winds is performed for this wind speed at the reference height of 10 m above the ground. The results of the CFD analysis are assessed again in the horizontal plane, placed 1.75 m above the ground (Fig. 2.12). The three buildings influence the character of the flow that approaches the

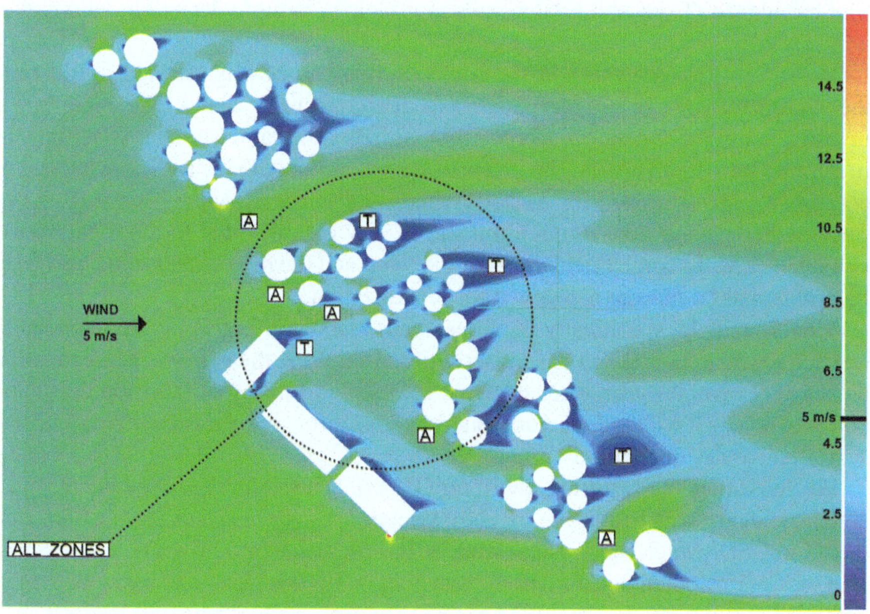


**Fig. 2.11** Westerly wind flow through the silo clusters. The area of interest is in the dotted circle

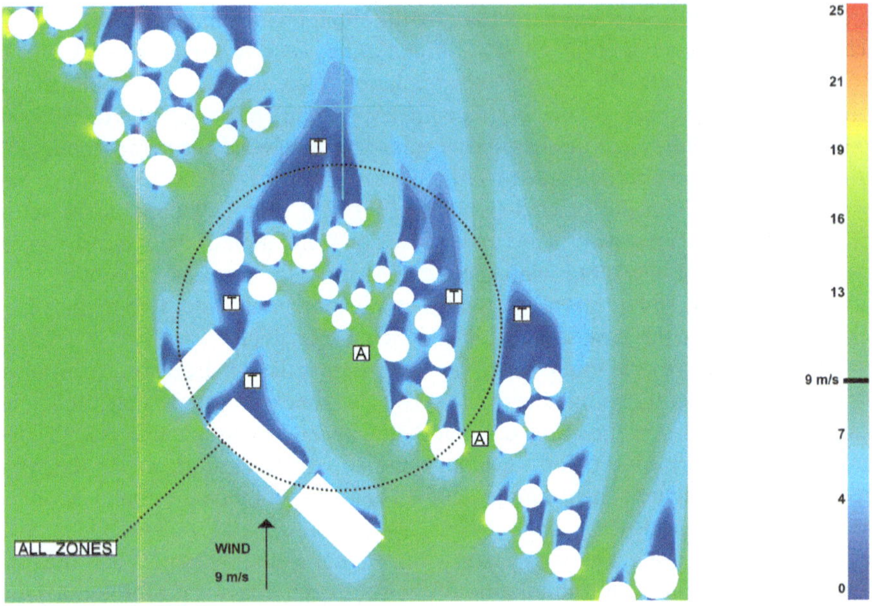


**Fig. 2.12** Southerly gusts through the silo clusters. The area of interest is in the dotted circle

silos. The accelerated wind (A) reaches speeds from 13 to 19 m/s (green to yellow). The turbulence (T) caused by the site's morphology is more detectible with higher wind speeds.

Because of the large area of the docks, only a part of it is selected to illustrate the wind-driven design and morphogenesis of architectural shapes. The wind flow is investigated in more detail in Rhino CFD, utilizing the same simulation settings as in the previous simulations. The zones of turbulent flow (T), as well as wind acceleration (A), are examined for westerly winds and southerly wind gusts (Fig. 2.13). The southerly wind flow through the site is very complex due to the higher wind velocities. The dynamic fluctuations of the wind and the momentary wind gusts can be problematic if the site is planned for revitalization.

The brownfield, specifically the three zones marked in Fig. 2.13, are intended to be used for recreational and leisure activities. Therefore, the extreme wind situation during 18.77% of the year, when the wind velocity is >5 m/s, should be addressed and possibly reversed with architectural interventions through the proposed wind-driven performative design method.

For subsequent CFD testing, Swift for GH will be used for the following reasons: (i) the geometry with negligible (or zero) thickness can be tested in Swift, contrasted to Rhino CFD, which requires meshing of the geometry into 3D cells, hence a fine grid of VWT calculation cells, (ii) the meshing in Swift is faster and easier to set-up, and (iii) Swift utilizes the OpenFOAM CFD platform, with comparable performance as Ansys Fluent [17].

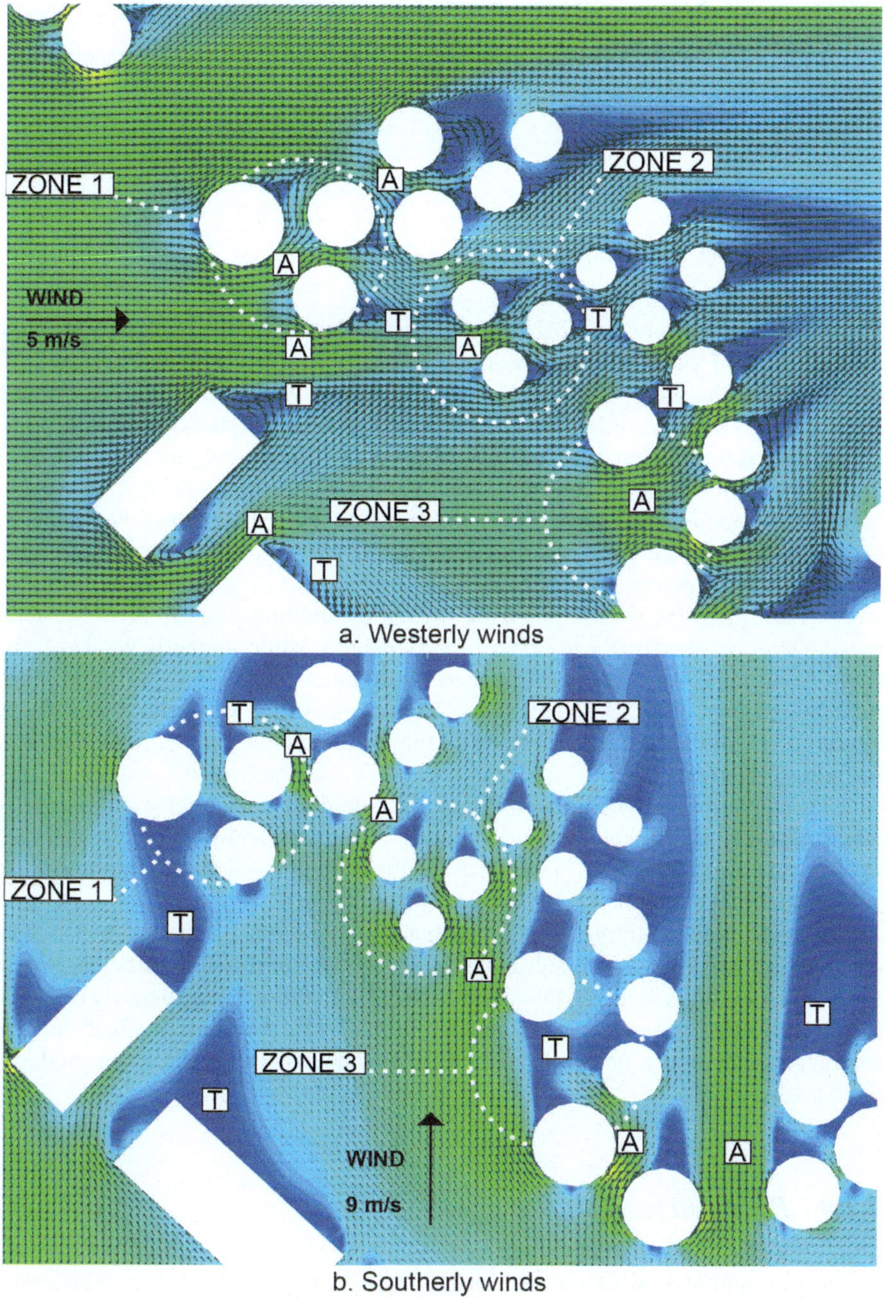


**Fig. 2.13** Close-up of the area of interest in westerly winds (top), and southerly winds (bottom)

**Table 2.3** VWT settings for the tests of the design site with modified silo heights

| Cell size (m) | xy-scale | xy-offset | z-scale | Terrain roughness | Relaxation factors | Convergence criteria |
|---|---|---|---|---|---|---|
| 4 | 1.1 | 20 | 6 | 1 | $p = 0.3$<br>$v_{x,y,z} = 0.7$<br>$k|\varepsilon = 0.7$ | $p = 5e^{-2}$<br>$v_{x,y,z} = 1e^{-2}$<br>$k|\varepsilon = 1e^{-2}$ |

### 2.2.3 *Urbanism: Design*

The idea is to convert the abandoned brownfield into a 'wind park' filled with recreational and cultural activities. Three different architectural interventions will uniquely alter the wind flow in each one of the three zones. In zone 1, swimming pools (or ice rinks in winter) will be proposed, designed to deflect the wind. The silos intended for this function are reduced in height, leaving only the bottom ring of the concrete silo. Zone 2 works with wind acceleration; it concentrates the wind to harvest the wind energy through piezoelectric cantilevered elements on the tensile membrane wrapping the silos. In zone 3, the aerodynamic shape is proposed, serving as a cultural auditorium.

After adjusting the silos heights, the site wind conditions are explored. Westerly winds are represented by the inlet wind speed of 5 m/s, while southerly wind gusts by 9 m/s. Swift for Grasshopper is utilized to investigate the wind flow passing through the three zones. Table 2.3 depicts the VWT settings of the westerly and southerly winds CFD simulations. The meshing of the virtual wind tunnel is refined around the area of interest.

The calculations converged in 495 and 388 iterations for westerly and southerly winds, respectively. Accelerated wind flow (A) and turbulent flow (T) are depicted in Figs. 2.14 and 2.15. The CFD analysis results are displayed on a horizontal plane, 1.75 m above the ground.

The wind flow, colliding with the rectangular buildings on the southwest side, deflects from this blockage, is subsequently accelerated, and creates turbulent wakes. In zones 1 and 3, this dynamic flow is not wanted. Therefore, through newly designed architecture, it will be redirected. In zone 2, however, the Venturi effect[6] will be leveraged to harvest the energy from the accelerated wind, especially in the southerly wind gusts.

### 2.2.4 *Parametric Design*

Twining around and in-between the cylindrical silos, three innovative architectural shapes work with the wind in the unique morphological conditions (Fig. 2.16). The term 'FlowBrane' is created in this work, standing for a tensile membrane that

[6] Venturi effect—when a fluid passes through a constriction, there is an increase in its velocity.

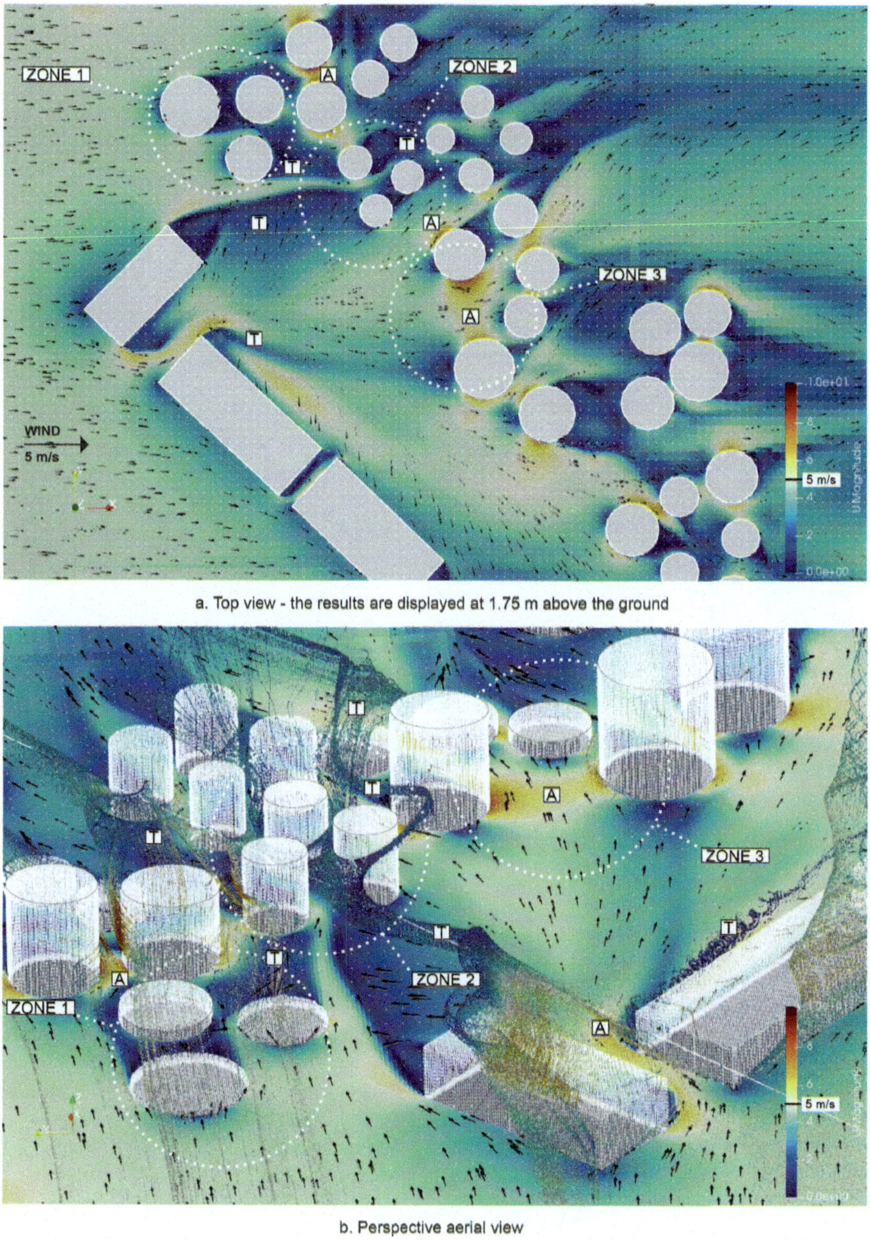


**Fig. 2.14** Westerly winds CFD analysis—the silo heights are regulated according to the proposal. Zone 1 is planned for baths; zone 2 is focused on the wind energy harvesting, while a cultural auditorium is proposed in zone 3

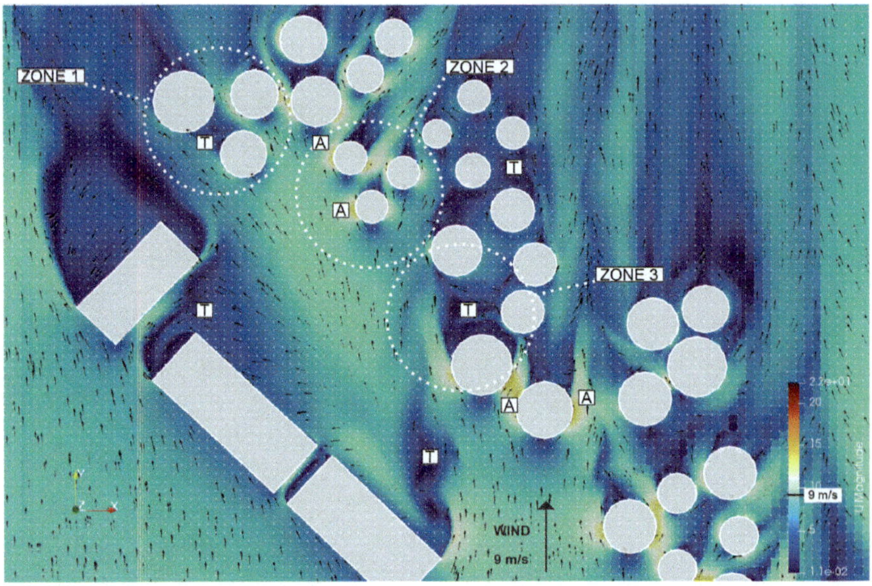


a. Top view - the results are displayed at 1.75 m above the ground

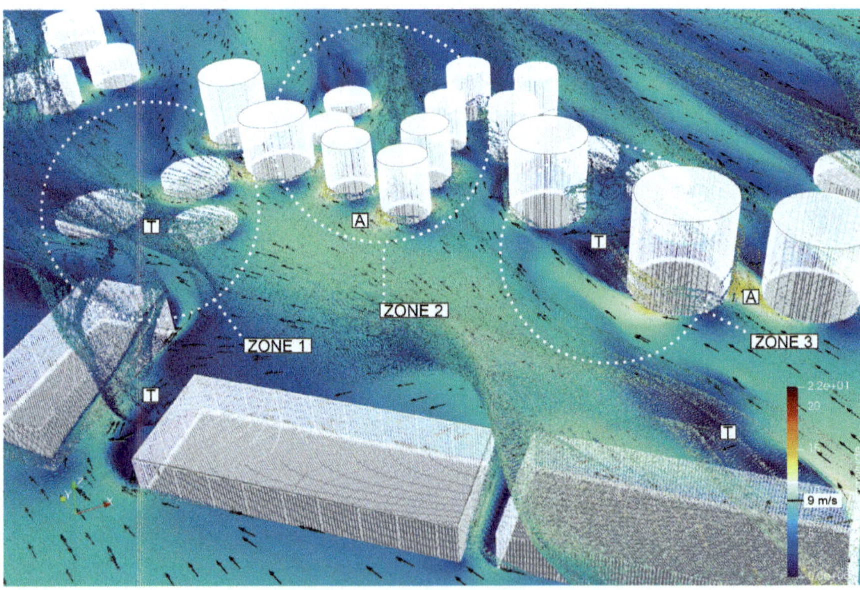


b. Perspective aerial view

**Fig. 2.15** Southerly winds analysis—the silo heights are regulated according to the proposal. Zone 1 is planned for baths; zone 2 is focused on the wind energy harvesting, while a cultural auditorium is intended in zone 3

**Fig. 2.16** The 'wind park' with FlowBrane design

controls wind flow. The parametrically-created shapes can be smoothly modified, contributing to faster CFD testing.

#### 2.2.4.1 Deflecting the Wind Flow in Zone 1

The FlowBrane no. 1, designed as a windbreak from the prevailing westerly, and strong southerly winds, is intended to deflect the flow in both the horizontal and vertical direction. A lightweight, porous tensile membrane, vertically inclined towards the flow, is created to protect the baths and the sitting zones against the cold Swedish winds while transmitting light and heat from the sun (Fig. 2.17). Medium-permeable barriers (>30% porosity) are usually more effective in reducing the wind speed and turbulent wind energy, as well as increasing the length of the sheltered space on the leeward side than solid windbreaks. The oblique wind approaching angle from about 45° decreases the effectiveness of windbreaks [18, 19].

For this reason, the intended membrane is designed as a wavy shape, suitable for different wind-approaching angles. Although partly permeable, the membrane will be tested in the CFD analysis as a solid shape.

The two parametric variables, the height $h$ and the lift angle $\alpha$ of the wind-shielding membrane are interdependent throughout the form-finding of FlowBrane no. 1. Another input variable is a 2D contour curve, through which the height of the membrane (constrained between 3 and 9 m) and its inclination to the approaching wind can be modified (Fig. 2.18).

#### 2.2.4.2 Accelerating the Wind Flow in Zone 2

Wind acceleration caused by the original morphology of the silo structures in Zone 2 is not suppressed by design. On the contrary, its effects are enhanced. Employing

**Fig. 2.17** The visualization of FlowBrane no. 1, protecting the baths. The emptied silos, reduced in their height, are now filled with water, and serve for swimming activities

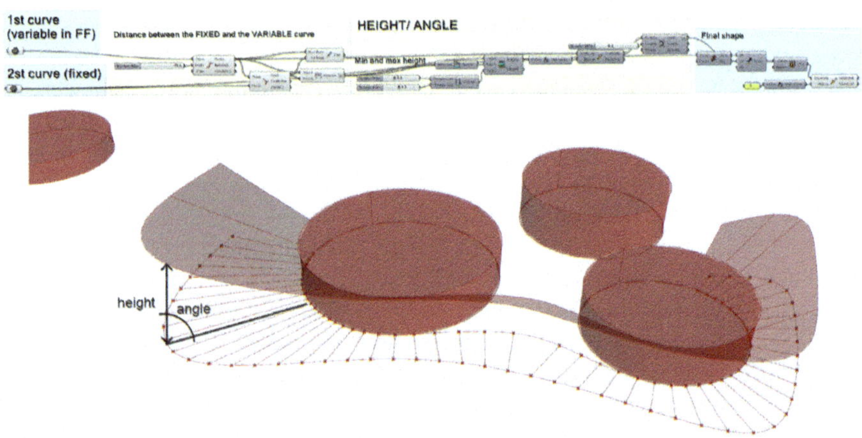


**Fig. 2.18** The Grasshopper definition, controlling the final shape of FlowBrane no. 1

the specific shape of FlowBrane no. 2, designed to twine around three high silos, the wind is captured and squeezed between the membranes utilizing the Venturi effect and directed in the desired way. Fluttering in the wind, the cantilevered, 1.2 m long piezoelectric straws can efficiently harvest wind energy, particularly in wind gusts (Fig. 2.19). When the resonant frequency of the piezoelectric harvester matches the wind vortex shedding frequency, high amounts of energy can be generated [20].

Employing Grasshopper, a set of three membranes is designed. The bottom contour curve is created from four points around the three silos. By rotating and lifting

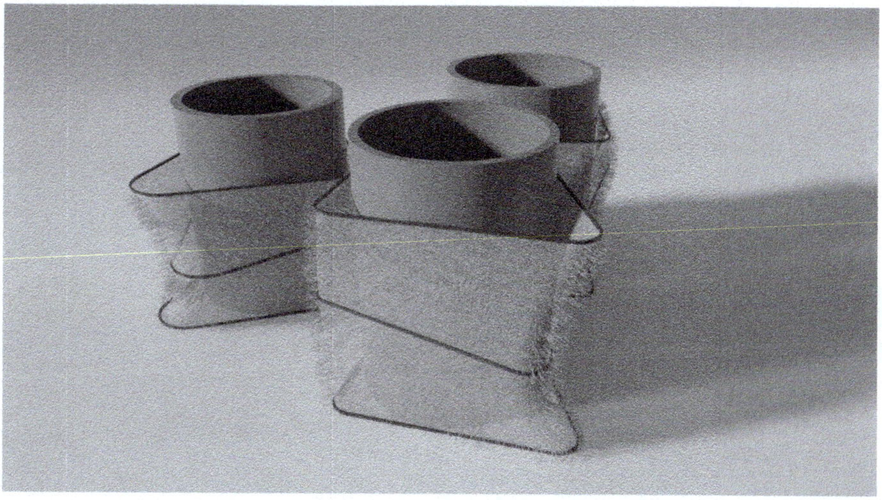

Fig. 2.19 FlowBrane no. 2 with cantilevered piezoelectric straws

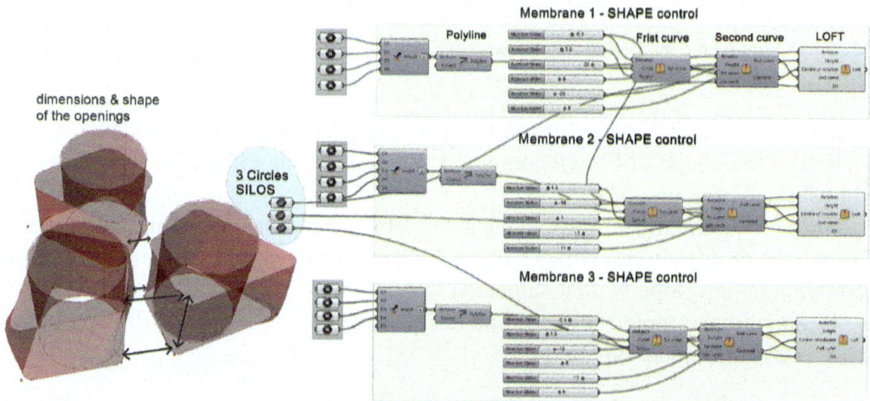


Fig. 2.20 The Grasshopper definition, controlling the final shape of FlowBrane no. 2

this curve, the other two contours of FlowBrane no. 2 are subsequently formed. The contour curves control the global shapes and hence the size and form of the openings between each membrane, which, consequently, affects the wind fluxes (Fig. 2.20).

#### 2.2.4.3 Minimum Resistance to the Wind Flow in Zone 3

The intended function of FlowBrane no. 3 is a cultural auditorium between two silos. The goal is to create an aerodynamic form, resulting in a minimum resistance in the wind flow, thus almost no turbulence (Fig. 2.21).

**Fig. 2.21** FlowBrane no. 3 with a wind-adaptive building envelope

A wind-adaptive building envelope, introduced in the following chapter, will be applied to the shape of FlowBrane no. 3 to examine the effects on wind flow and wind surface pressure.

Through parametric definition, the shape of the third tensile membrane is created through the profile curves, which control the length and height of the auditorium. The aerodynamic envelope is then subjected to wind pressure (Fig. 2.22), transforming the façade into a dimpled surface. The next chapter will explain how this influences the overall building aerodynamics.

### *2.2.5 CFD Wind Analysis Through Swift for Grasshopper*

Leveraging the parametric approach, we can create multiple shape variants for the subsequent wind testing. Only one shape option for each FlowBrane, representing a different category of wind-architecture interactions, is selected, and examined here.

First, the whole site is analyzed with the three designed FlowBrane shapes. The VWT settings are the same as in the CFD tests with adjusted silo heights (Table 2.3). Despite the initial cell size being as large as 4 m, the meshing is more refined around the area of interest. The horizontal plane for displaying the results is placed 1.75 m above the ground (Figs. 2.23 and 2.24). That implies that the wind situation is captured at the same height as in the simulations without architectural interventions.

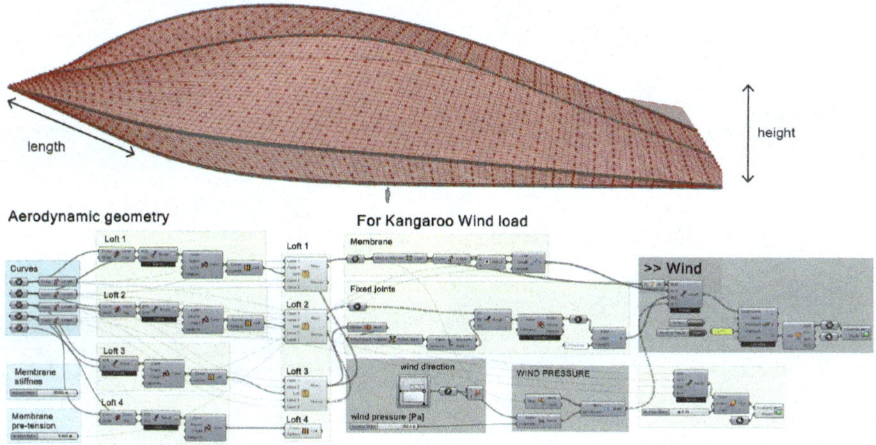


**Fig. 2.22** The Grasshopper definition, controlling the final shape of FlowBrane no. 3

Nevertheless, these CFD tests provide a basic idea of the impact of all designed shapes on the wind flow, juxtaposed to the situation before architectural interventions. The CFD results, presented in the perspective view, help understand the differences.

Second, The CFD simulations of each designed shape are carried out separately to observe the wind-architecture interactions in more detail. The VWT settings and the convergence criteria are depicted in Table 2.4. For faster convergence, less strict convergence criteria for the pressure p are used in the simulations of FlowBrane 2 and 3.

The results demonstrate that the expected wind-architecture interactions in the three tested zones are successfully achieved. FlowBrane no. 1 deflects the flow, FlowBrane no. 2 concentrates it, while FlowBrane no. 3 minimally interferes with the wind. The subsequent sub-chapters will describe the results in more detail, individually for each designed shape.

#### 2.2.5.1 Deflecting the Wind Flow in Zone 1

The results of the CFD simulations confirm that FlowBrane no. 1, designed around the baths to protect pedestrians against the prevailing westerly winds and cold southerly wind gusts, meets the defined function.

##### Westerly Winds

The wind speed of 5 m/s is the input speed in the CFD simulations of the westerly winds (Fig. 2.25). The horizontal plane for displaying the CFD results is placed near the water surface of the swimming pools. Through deflecting the wind, the proposed

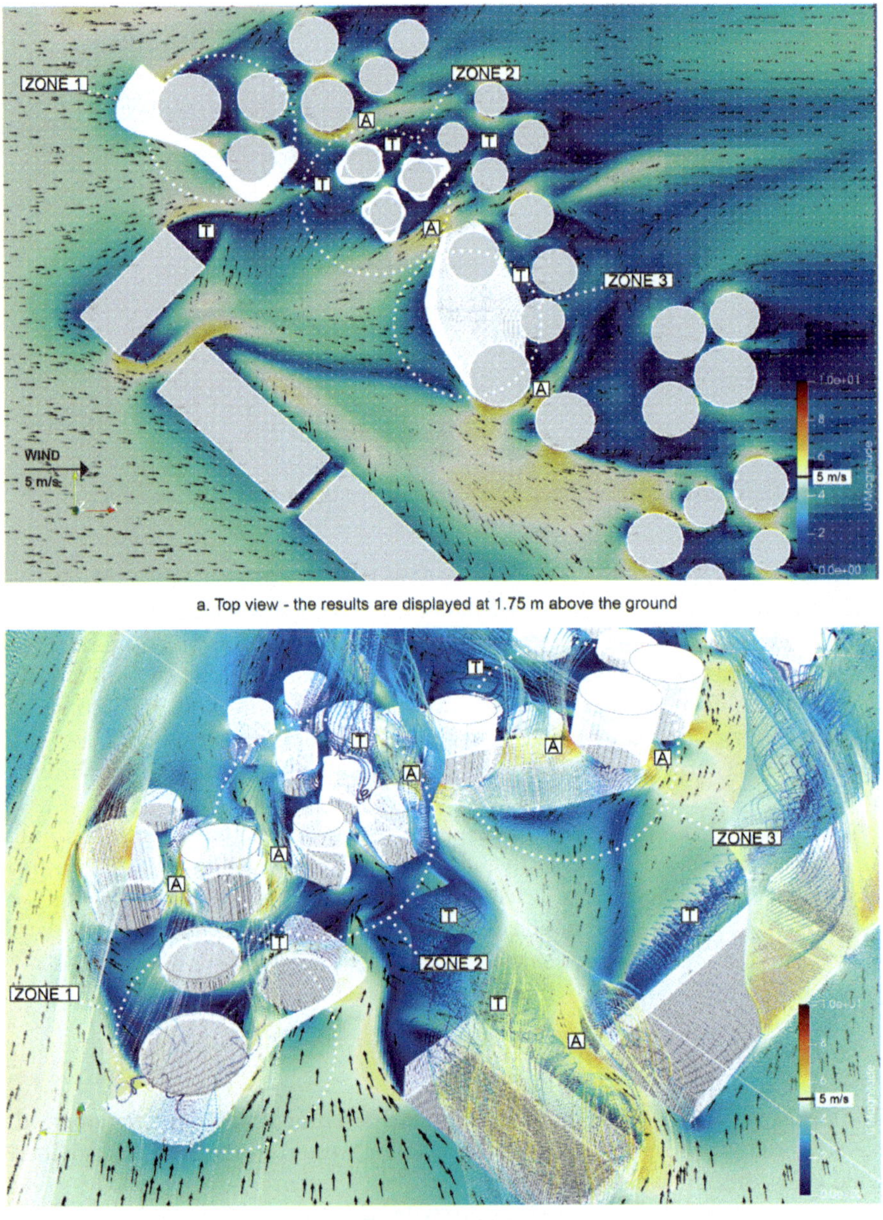


**Fig. 2.23** Westerly winds through the area of interest with the designed architectural forms. FlowBrane 1 in zone 1 deflects the wind from the baths. FlowBrane 2 in zone 2 accelerates the flow for wind energy harvesting. FlowBrane 3 in zone 3 creates an aerodynamically-shaped auditorium

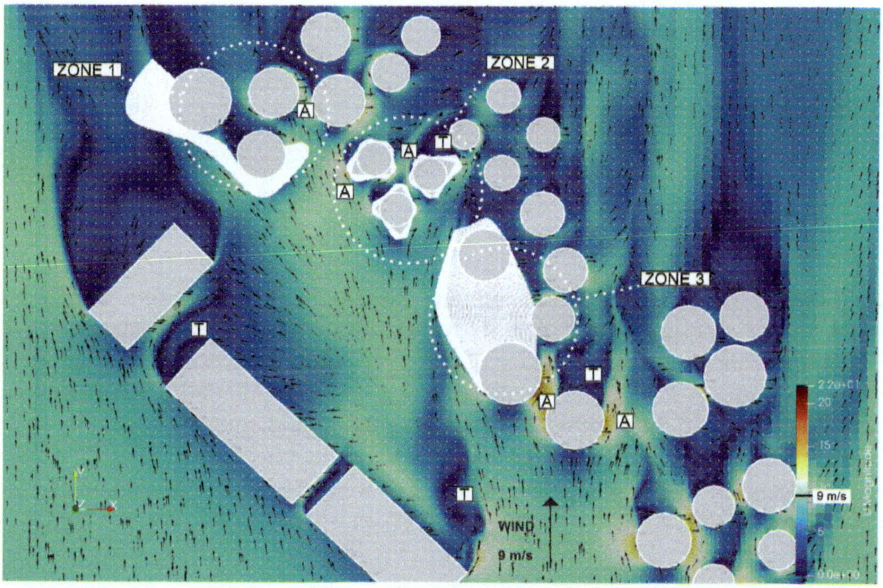


a. Top view - the results are displayed at 1.75 m above the ground

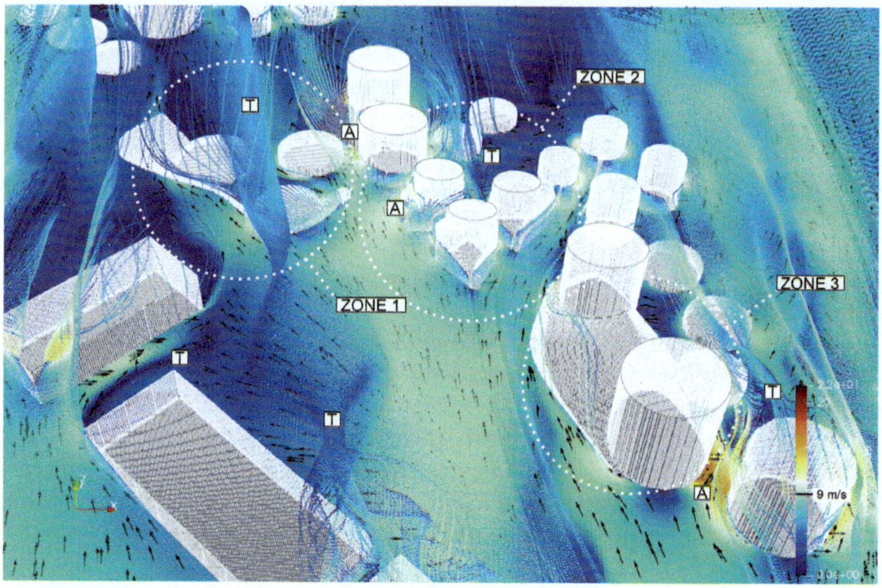


b. Perspective aerial view

**Fig. 2.24** Southerly wind gusts through the area of interest with the designed architectural forms. FlowBrane 1 in zone 1 deflects the wind from the baths. FlowBrane 2 in zone 2 accelerates the flow for wind energy harvesting. FlowBrane 3 in zone 3 creates an aerodynamically-shaped auditorium

**Table 2.4** VWT settings for the simulations of the FlowBrane 1, 2 and 3

| Cell size (m) | xy-scale | xy-offset | z-scale | Terrain roughness | Relaxation factors | Convergence criteria |
|---|---|---|---|---|---|---|
| 2 | 1.5 | 100 | 6 | 1 | $p = 0.3$<br>$v_{x,y,z} = 0.7$<br>$k\|\varepsilon = 0.7$ | $p =$ $4e^{-2}\|4.5e^{-2}$<br>$v_{x,y,z} = 1e^{-2}$<br>$k\|\varepsilon = 1e^{-2}$ |

lightweight, transparent tensile membrane can provide wind comfort, even at the wind comfort threshold speed. A weak wind swirl with a reduced wind speed of 0–3 m/s forms above the water surface of the bigger swimming pool. The wind speed above the water surface of the smaller swimming pool is about 0–2 m/s. The wind accelerates back to the inlet wind speed (5 m/s) on the membrane edges. It can be observed in the section through the smaller swimming pool (Fig. 2.25).

The designed tensile membrane in the 5 m/s winds is exposed to a minimal wind suction (Fig. 2.26), around −15 to −5 $m^2/s^2$ (i.e., −18.75 to −6.25 Pa).

Southerly Winds

In the southerly wind gusts, represented by the speed of 9 m/s, FlowBrane no. 1 effectively shields the swimming baths from the strong wind. The wind protection of the bigger pool is surprisingly good, despite the high wind velocity. The speed of the wind swirls, occurring above the water surface of the smaller pool, is around 0–6 m/s, though most of the area is calm. Zones of wind acceleration form in-between the swimming pools and on the outer edges of the membrane windbreak (Fig. 2.27).

In some regions of the tensile membrane, the wind suction can reach approximately −40 $m^2/s^2$ (i.e., −50 Pa). The distribution of wind surface pressure is depicted in Fig. 2.28.

The parametrically-generated wind-protecting tensile membrane, analyzed through Swift for GH, has a promising wind-deflecting function. The membrane is designed for two wind directions, providing wind comfort in the prevailing westerly winds while performing as a good wind deflector in the southerly gusts. In the search for the optimal shape of FlowBrane no. 1, the part in-between the two swimming pools could be lifted higher, or the inclination of the membrane to the approaching wind could be adjusted to reduce the wind acceleration between the silos.

#### 2.2.5.2 Accelerating the Wind Flow in Zone 2

Three silos in zone 2 induce the Venturi effect, as was determined through the CFD tests of the site with no architectural interventions. FlowBrane no. 2 is a set of membranes that supports and enhances the already occurring wind acceleration and leverages it to harvest the energy from the concentrated wind flow.

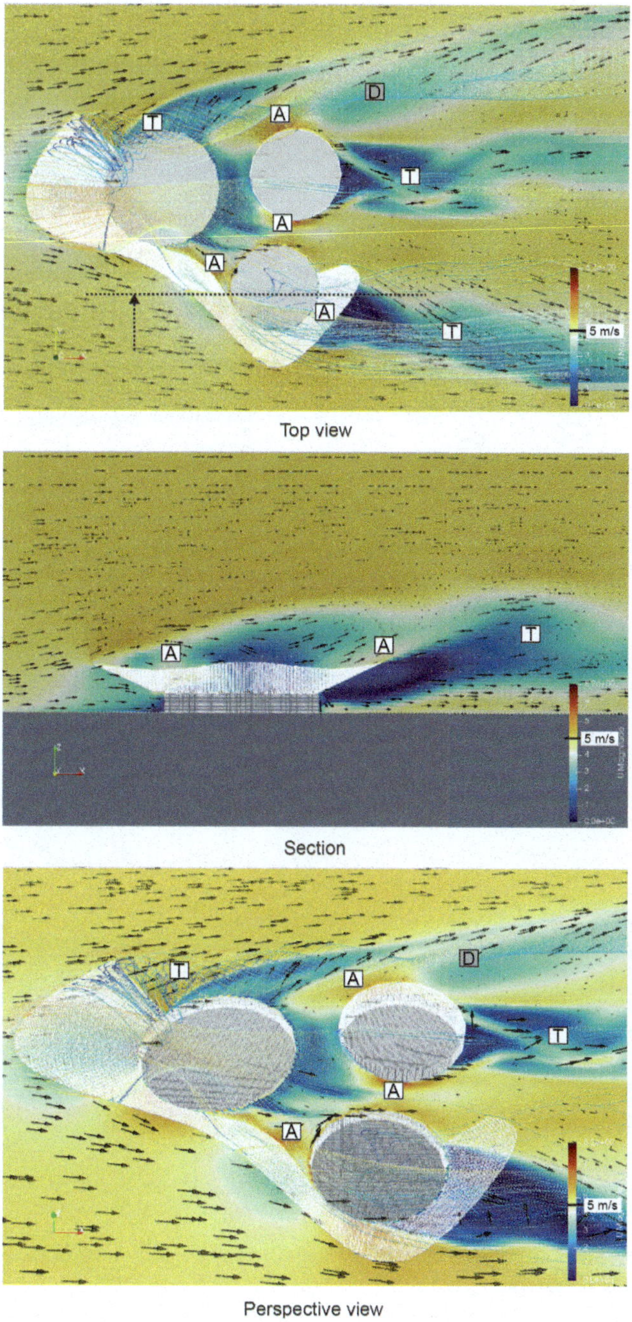


**Fig. 2.25** FlowBrane no. 1 in the westerly winds. (A) accelerated, (T) turbulent, (D) diffused flow

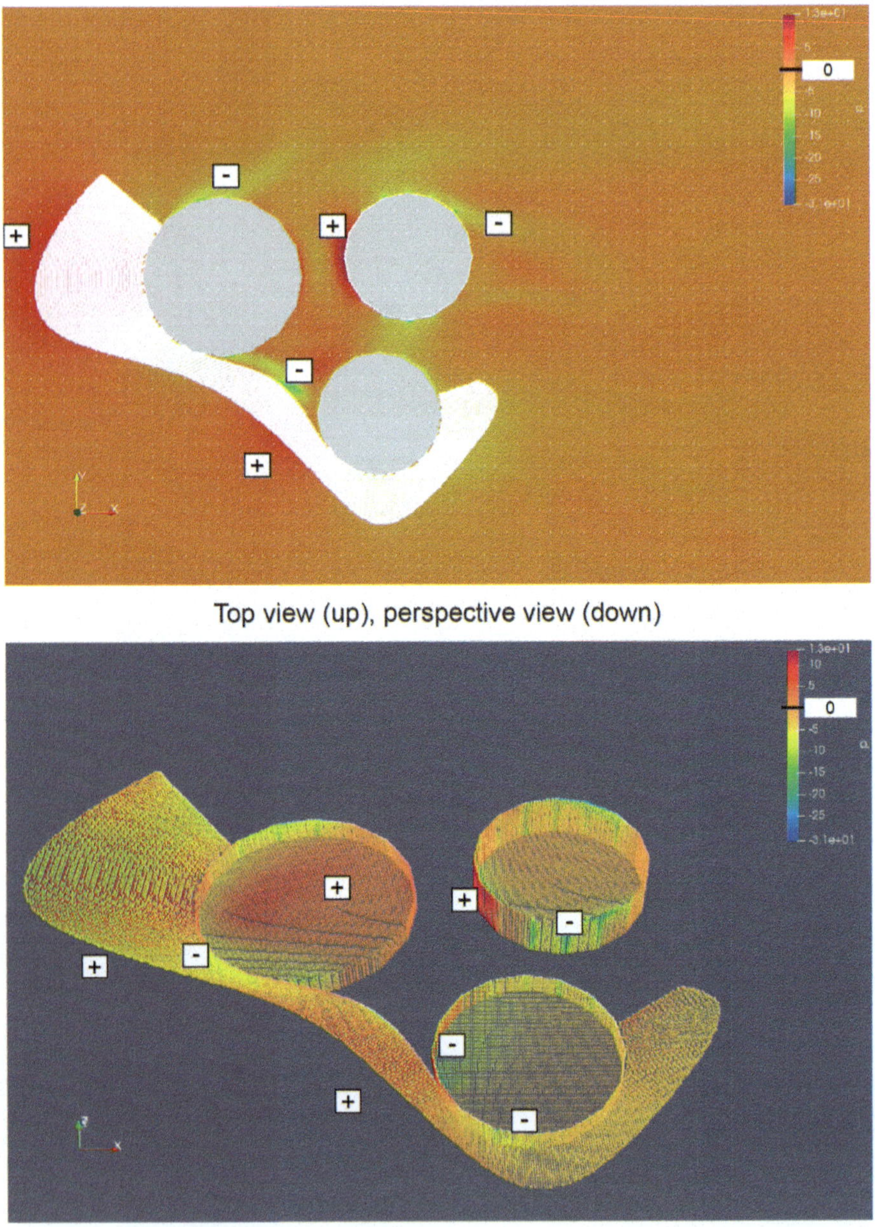


**Fig. 2.26** FlowBrane no. 1: the wind pressure in the westerly winds. Areas of the positive (+) and negative (−) wind pressure

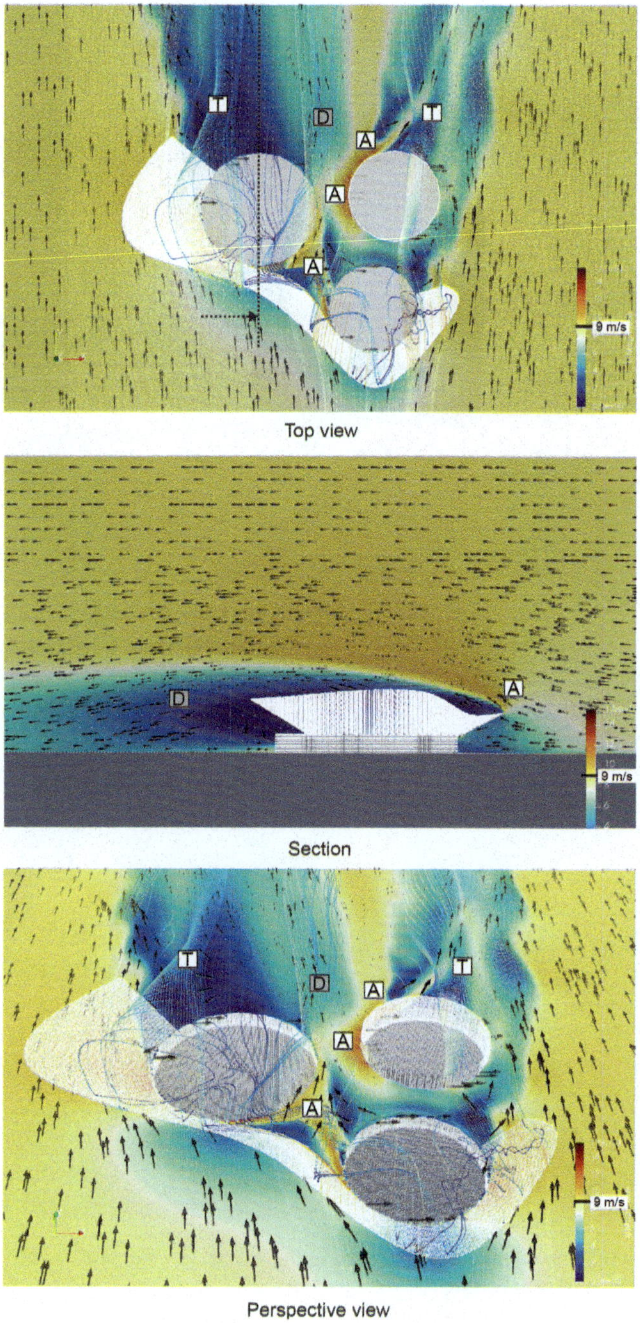


**Fig. 2.27** FlowBrane no. 1 in the southerly wind gusts. (A) accelerated, (T) turbulent, (D) diffused flow

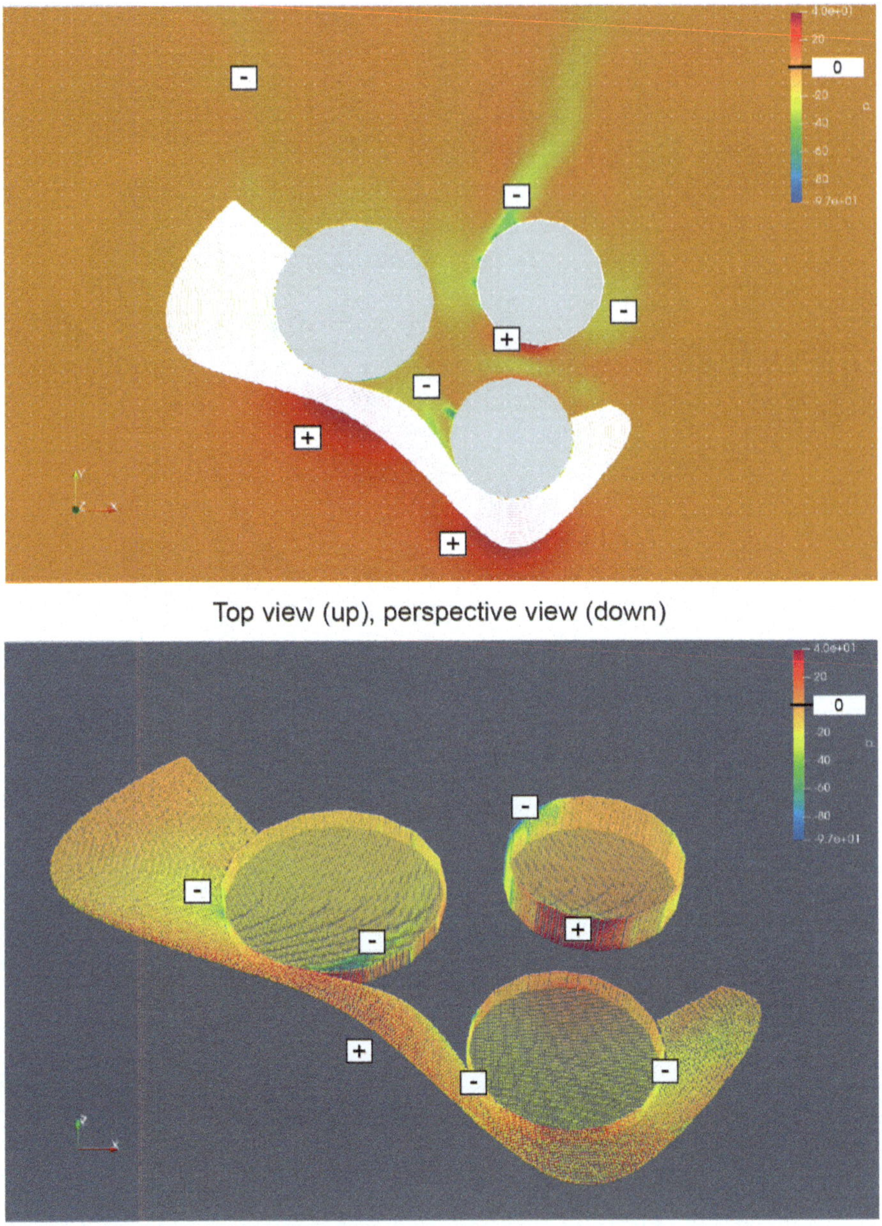


**Fig. 2.28** FlowBrane no. 1: the wind pressure in the southerly wind gusts. Areas of the positive (+) and negative (−) wind pressure

### Westerly Winds

The westerly winds CFD simulation demonstrates that even in the 5 m/s winds, the flow can be squeezed and attain the velocities between 6 and 9 m/s. Compared to the situation without the FlowBrane no. 2, higher wind acceleration is achieved by employing designed shapes (Fig. 2.29).

Downwind of FlowBrane no. 2, the flow is again diffused, sparsely creating a faint wind swirl, where the velocity close to zero prevails.

The accelerated flow creates a pressure difference between the zones upwind and downwind regarding the tested geometry (Fig. 2.30). On the lateral sides of the proposed FlowBrane no. 2, the wind suction reaches values of around −10 to −30 $m^2/s^2$ (i.e., −12.5 to −37.5 Pa). These areas of the tensile membrane are subjected to high stresses caused by wind acceleration; hence they are suitable for wind energy harvesting.

### Southerly Winds

The simulations in the southerly wind gusts, with the input wind speed of 9 m/s indicate that the wind, passing through the three membranes, can be accelerated to the twofold speed, ergo, up to 18 m/s (Fig. 2.31). With exiting the constricted space, the wind decelerates and is diffused while producing a turbulent wake, with wind velocity reaching up to 9 m/s. A calm zone is created downwind from the first silo impacted by the wind. In addition to the wind-accelerating function of the three designed tensile membranes, each individual silo-wrapping membrane is aerodynamically-shaped. This fact contributes to additional changes in the wind fluxes around and between FlowBrane no. 2.

Negative-pressure areas emerge on the leeward side of the tested geometry, simultaneously with the turbulence (Fig. 2.32). The lateral sides of the FlowBrane no. 2 are, in the southerly winds, subjected to wind suction reaching almost −120 $m^2/s^2$ (i.e., −150 Pa). The positive surface wind pressure on the windward side of the designed geometry is approximately +41 $m^2/s^2$ (i.e., +51.25 Pa).

The CFD simulations of the set of parametrically-generated tensile membranes aimed at accelerating the wind to harvest the energy demonstrated that the designed FlowBrane no. 2 meets the determined requirements. The wind acceleration is achieved in both the southerly wind gusts and the milder westerly winds.

With the piezoelectric elements (straws) fluttering in the wind, applied to the fluid shape of three lightweight membranes, the wind energy could be harvested. The idea of piezoelectric straws is not elaborated further, as the objective of the wind-driven approach is only to demonstrate the different wind-architecture interactions employed in the design process.

Parametric shape alterations of FlowBrane no. 2 can be focused on the turbulence reduction on the leeward side, as it decreases the efficiency of wind energy harvesting. That implies that the three shapes squeezing the wind in-between them should be designed yet more streamlined to induce less resistance to the wind.

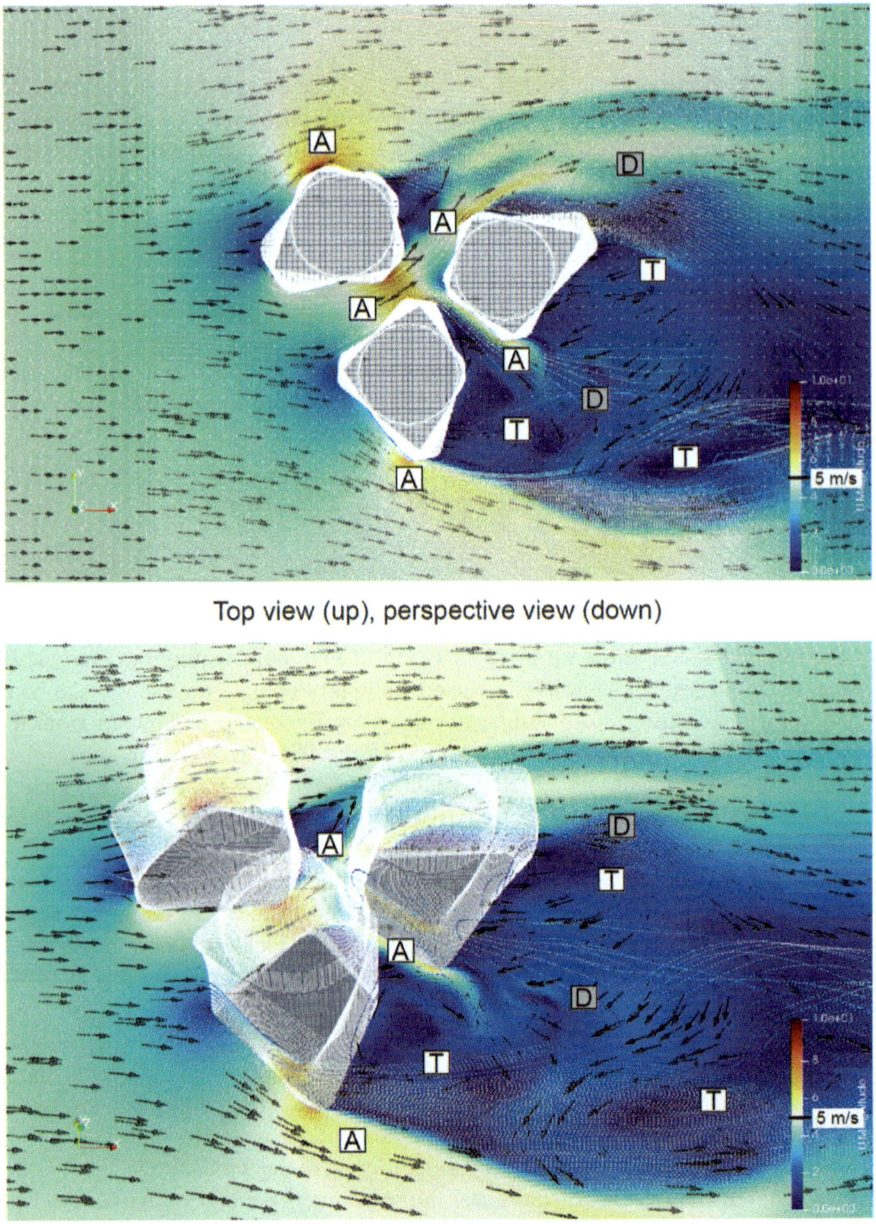


**Fig. 2.29** FlowBrane no. 2 in the westerly winds. (A) accelerated, (T) turbulent, (D) diffused flow

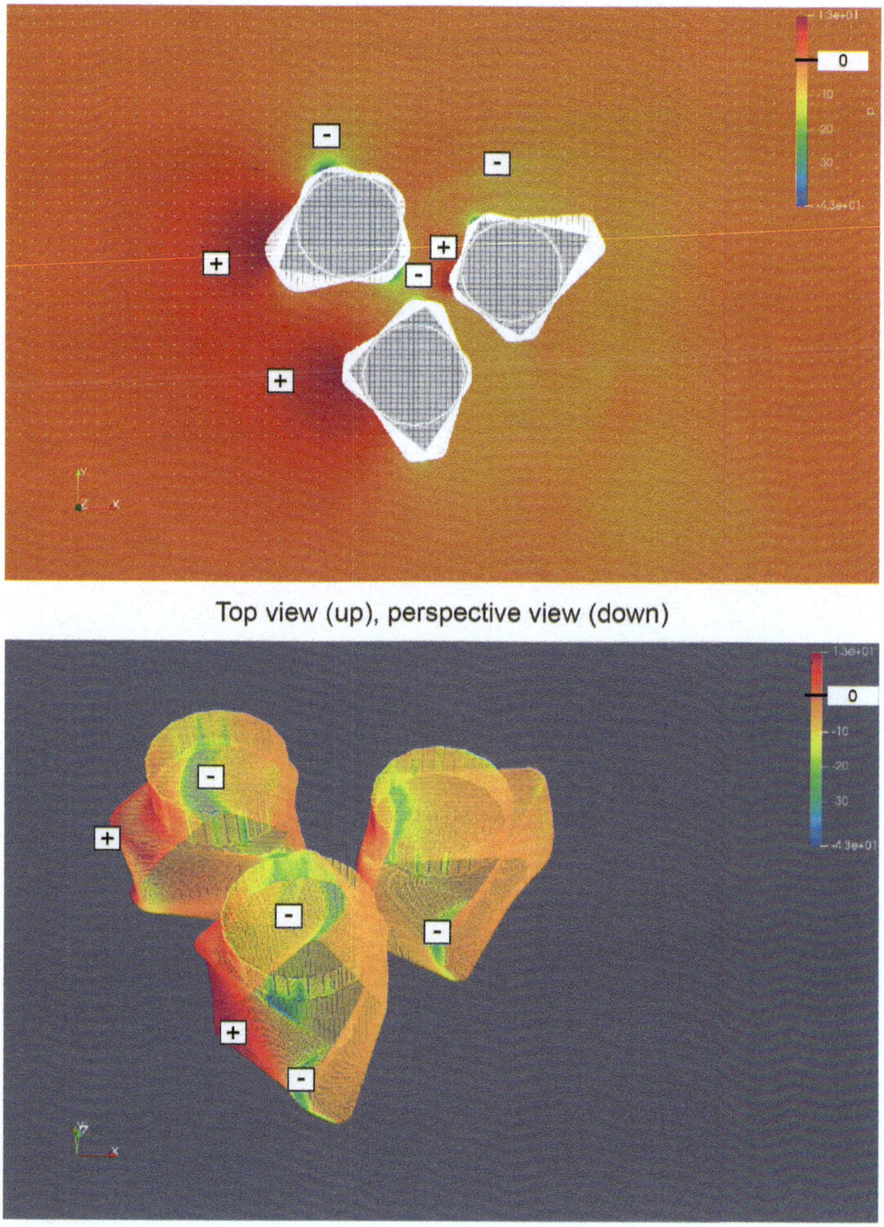


**Fig. 2.30** FlowBrane no. 2: the wind pressure in the westerly winds. Areas of the positive (+) and negative (−) wind pressure

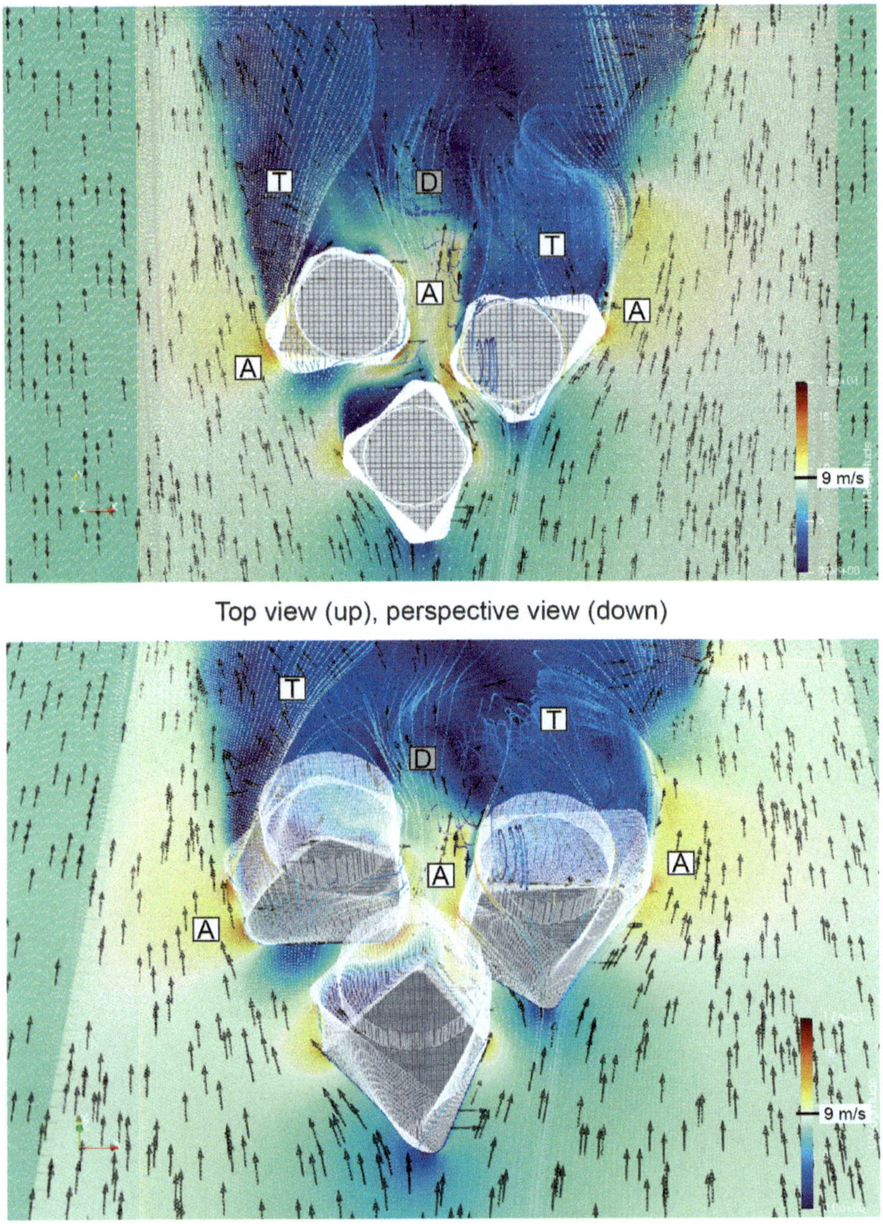


**Fig. 2.31** FlowBrane no. 2 in the southerly wind gusts. (A) accelerated, (T) turbulent, (D) diffused flow

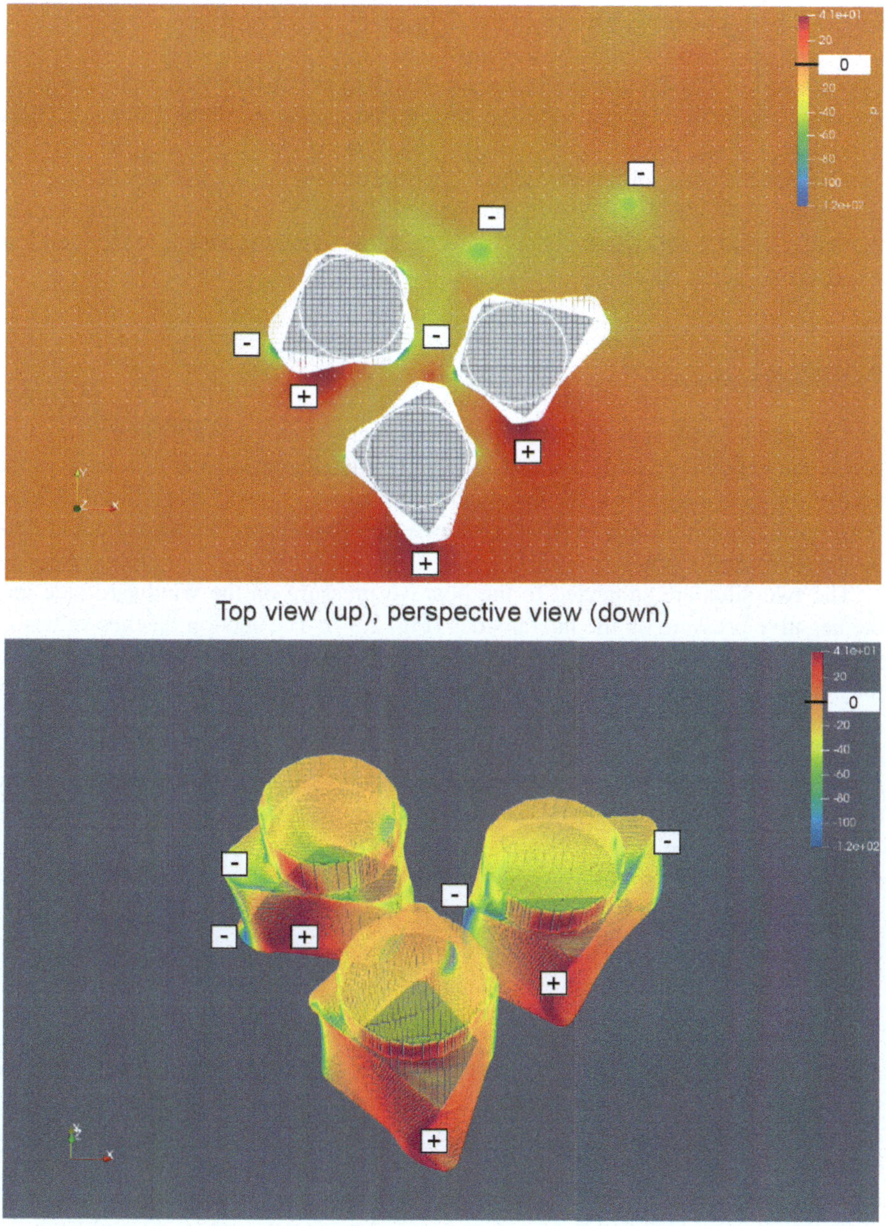


**Fig. 2.32** FlowBrane no. 2—the wind pressure in the southerly wind gusts. Areas of the positive (+) and negative (−) wind pressure

#### 2.2.5.3 Minimum Resistance to the Wind Flow in Zone 3

FlowBrane no. 3 is a cultural auditorium designed in-between two high silos as an aerodynamic form demonstrating the design that causes minimal resistance to the wind flow. The site's wind conditions after reducing the silo heights were investigated in the section Urbanism: Proposal. In the westerly winds, the zone between the two silos selected for the placement of the third membrane is exposed to the accelerated flow (Fig. 2.14). In the southerly gusts, on the other hand, the zone is in the turbulent wake of one of the silos (Fig. 2.15).

Westerly Winds

The designed aerodynamic form diverts the westerly winds to the sides, causing it to accelerate to approximately 6–7 m/s. The vertically-deflected wind flow induces the acceleration on the peak of the designed shape, close to one of the silos. The wind flow, hitting the smaller cylindrically-shaped silo, transforms into the turbulent flow on the silo's upper side (Fig. 2.33).

The two silos are subjected to the positive pressure on the windward side and the negative pressure on the lateral sides (Fig. 2.34). The suction reaches values of approximately $-28$ m$^2$/s$^2$ (i.e., $-35$ Pa). The windward side of the membrane, near the membrane-silo intersections, is subjected to the wind pressure of approximately $+17$ m$^2$/s$^2$ (i.e. $+21.25$ Pa). The wind suction on the peak of the membrane near the larger silo ranges between $-5$ and $-20$ m$^2$/s$^2$ (i.e., $-6.25$ to $-25$ Pa). Except for the local extremities, the designed shape is subjected to pressure of approximately 0 to $+7$ m$^2$/s$^2$ (i.e., 0 to $+8.75$ Pa). The drag coefficient $c_d$ of the designed shape in westerly winds equals 0.396, which indicates the shape can be more streamlined than a sphere (see section The wind loads in Chap. 1).

Southerly Winds

In the southerly 9 m/s wind gusts, the shape of FlowBrane no. 3 is well streamlined and generates almost no blockage to the wind flow. A lateral-side acceleration is produced by the silo on the windward side, causing a stream of air to hit the second silo and create a turbulent wake on the leeward side. The wind speed in the wake zone ranges between 0 and 8 m/s. The flow on the right side of the upwind silo accelerates up to the velocity of 17 m/s (Fig. 2.35).

Both silos are subjected to a positive wind pressure of around $+43$ m$^2$/s$^2$ (i.e., $+$ 53.75 Pa) on the windward side (Fig. 2.36). The lateral sides of silos are subjected to the wind suction, reaching $-120$ m$^2$/s$^2$ (i.e., $-150$ Pa). A wind suction zone of approximately $-60$ m$^2$/s$^2$ (i.e., $-75$ Pa) is induced by the accelerating wind on the right side of the first wind impact silo. There are few wind suction areas on the surface of the designed membrane, where the suction reaches approximately $-40$ m$^2$/s$^2$ (i.e., $-50$ Pa). The positive wind pressure on the membrane surface can get to $+43$ m$^2$/s$^2$

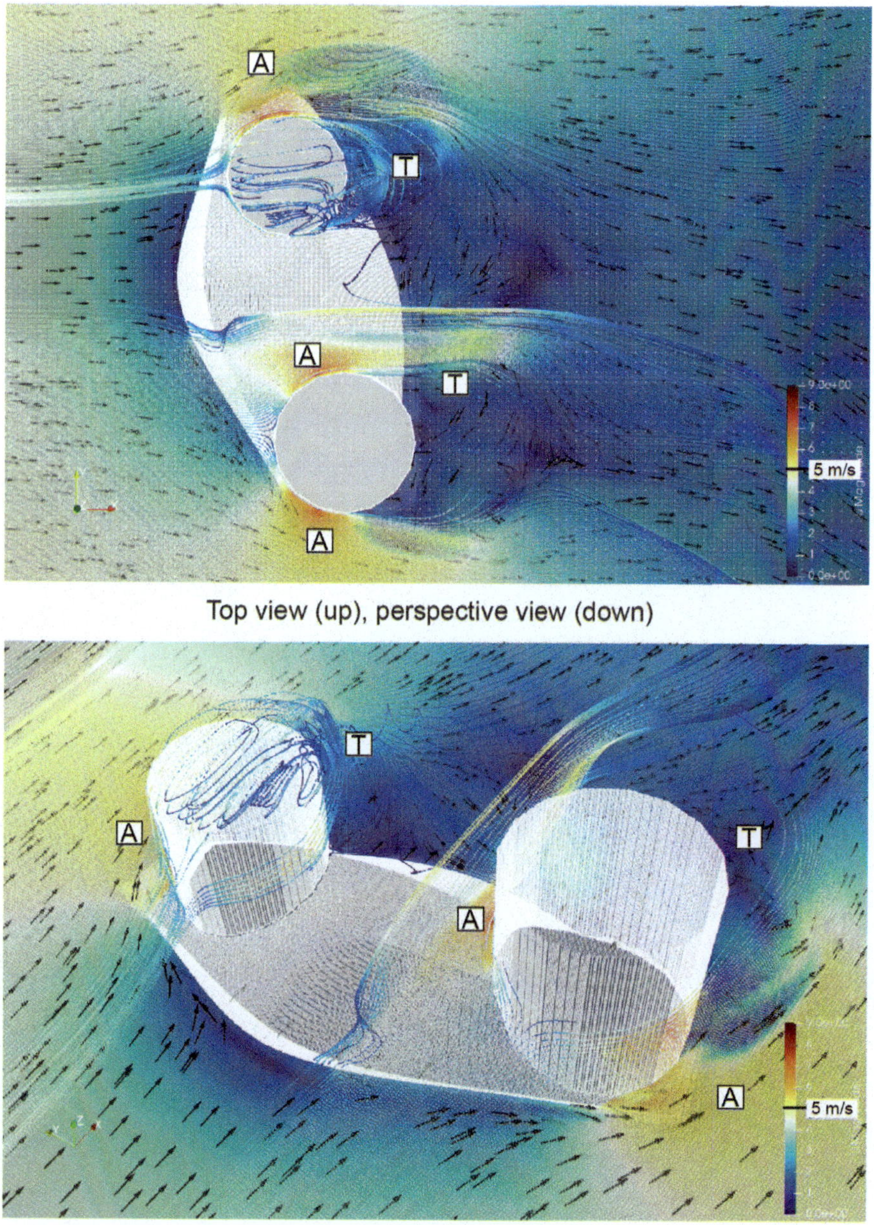


**Fig. 2.33** FlowBrane no. 3 in the westerly winds. (A) accelerated, (T) turbulent, (D) diffused flow

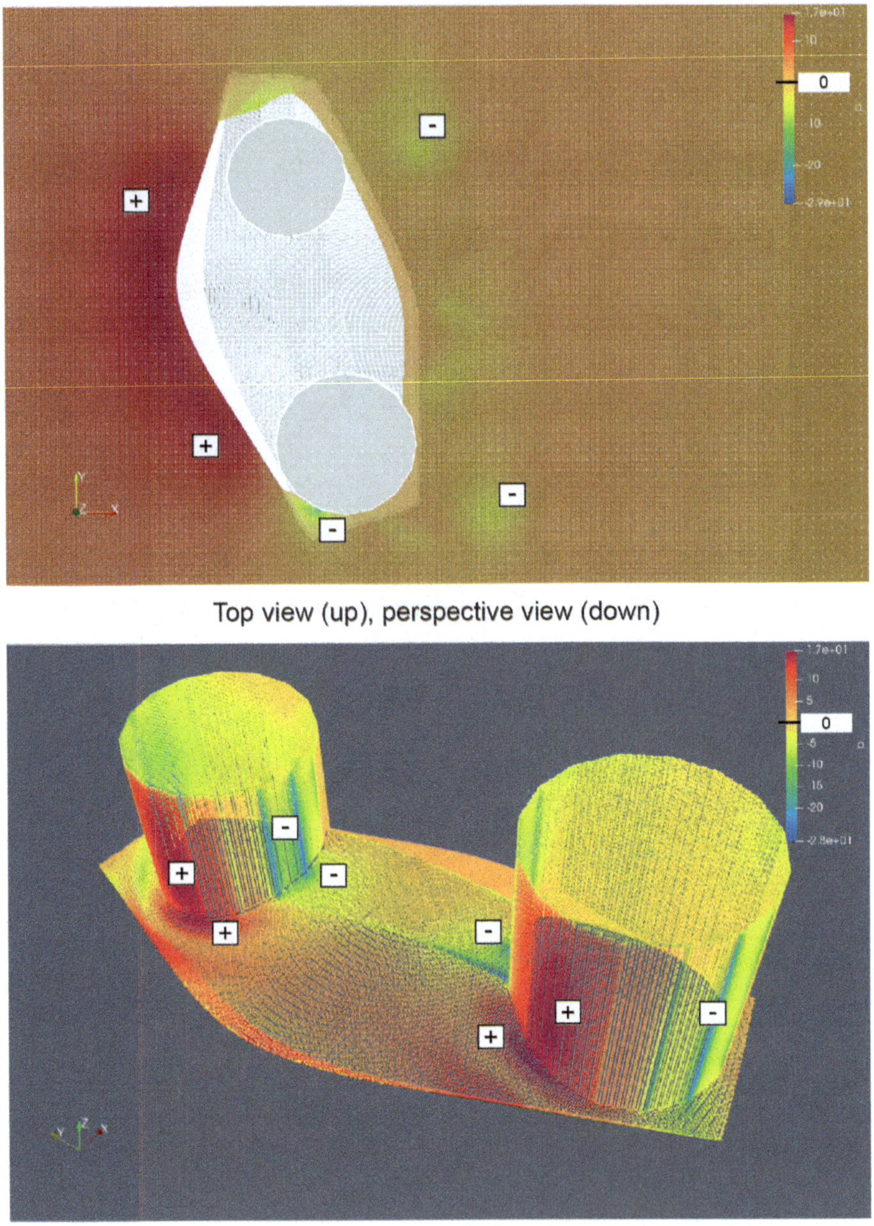


**Fig. 2.34** FlowBrane no. 3—the wind pressure in the westerly winds. Areas of the positive (+) and negative (−) wind pressure

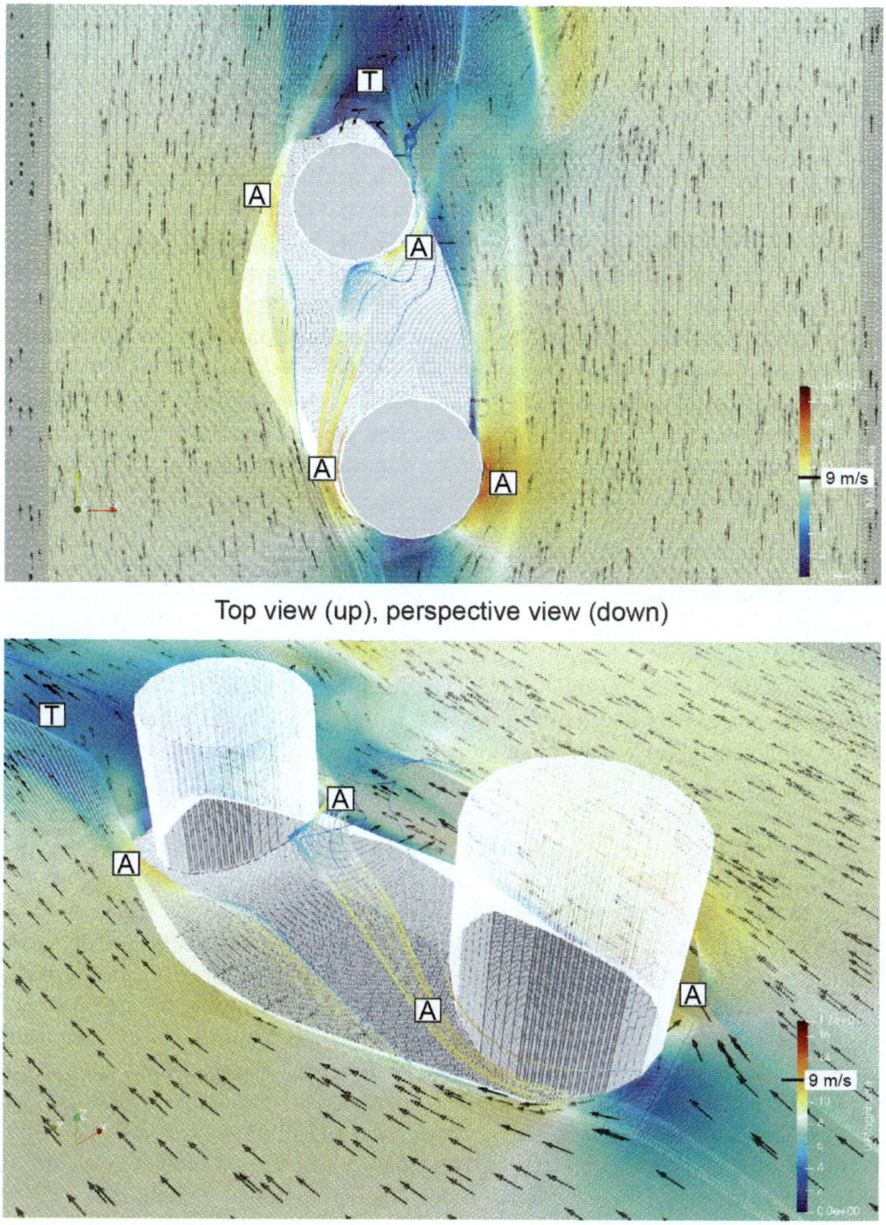


**Fig. 2.35** FlowBrane no. 3 in the southerly wind gusts. (A) accelerated, (T) turbulent, (D) diffused flow

(i.e., +53.75 Pa). The drag coefficient of the designed shape in the southerly winds is 0.049. This value is typical for a perfectly streamlined body (see Fig. 1.7).

The design goal when creating the FlowBrane no. 3 was to achieve a minimum resistance of the shape to the wind flow. The designed shape is more aerodynamic in the southerly wind gusts than in the westerly winds. The drag coefficient $c_d$, expressing the resistance to the wind flow, equals 0.049 in the 9 m/s southerly winds. In the westerly winds, the drag coefficient equals 0.396. The designed envelope has a lightweight tensile membrane skin. Because of its aerodynamic form, FlowBrane no. 3 is subjected to reduced surface pressure, positive or negative, induced by the wind flow.

The analyzed shape is intentionally more streamlined in the south-north than in the west–east direction, considering that the southerly winds are typically stronger in Stockholm, often accompanied by wind gusts. Further optimization of the shape through the parametric approach can be focused on redesigning the west façade of the auditorium, so it is even more fluid. Lower wind pressure on building surfaces can lead to lighter building envelopes.

#### 2.2.5.4 Conclusions

This book chapter introduced an architectural transformation of Loudden Docks' brownfield. In the unique wind microclimate, three FlowBrane shapes, designed as lightweight tensile membranes, interact with the wind and alter its flow. The reciprocal interplay of wind and architecture depends on the expected architectural outcome: (i) a deflection of the wind to protect swimming baths, (ii) acceleration of the flow to improve the efficiency of wind energy harvesting, or (iii) creating a minimum resistance to the wind, which leads to lighter building structures.

The parametric approach, coupled with fast parametric CFD simulations integrated into the working environment of Grasshopper, is a basis for the new design strategy of creating architectural shapes linked with nature. From the first idea, through wind-driven form-finding, followed by evaluating the wind performance using Swift (or other CFD plugin for Grasshopper), and looping back to the parametric alterations of the geometry, this method leads towards the environment-responsive architecture. Although Paraview, an external software for post-processing the CFD results, must be repetitively used for each tested shape variant, this is not perceived as a drawback. Besides, most of the CFD plugins offer a visual representation of the results within GH. This visual representation, however, cannot be graphically fine-tuned, hence it is not utilized here nor further elaborated.

If the whole architectural design process was VR-based, and the virtual reality was a natural part of designing, the evaluation of the CFD results using virtual reality would further challenge the boundaries of the environment-related designing.

In the day-to-day architectural practice, other wind analysis techniques can be leveraged for different design phases: (i) ML-based methods, or FFD can be engaged to analyze larger urban zones and get a quick idea of the wind flow character, (ii) Grasshopper CFD plugins can be utilized to analyze buildings on a smaller scale or

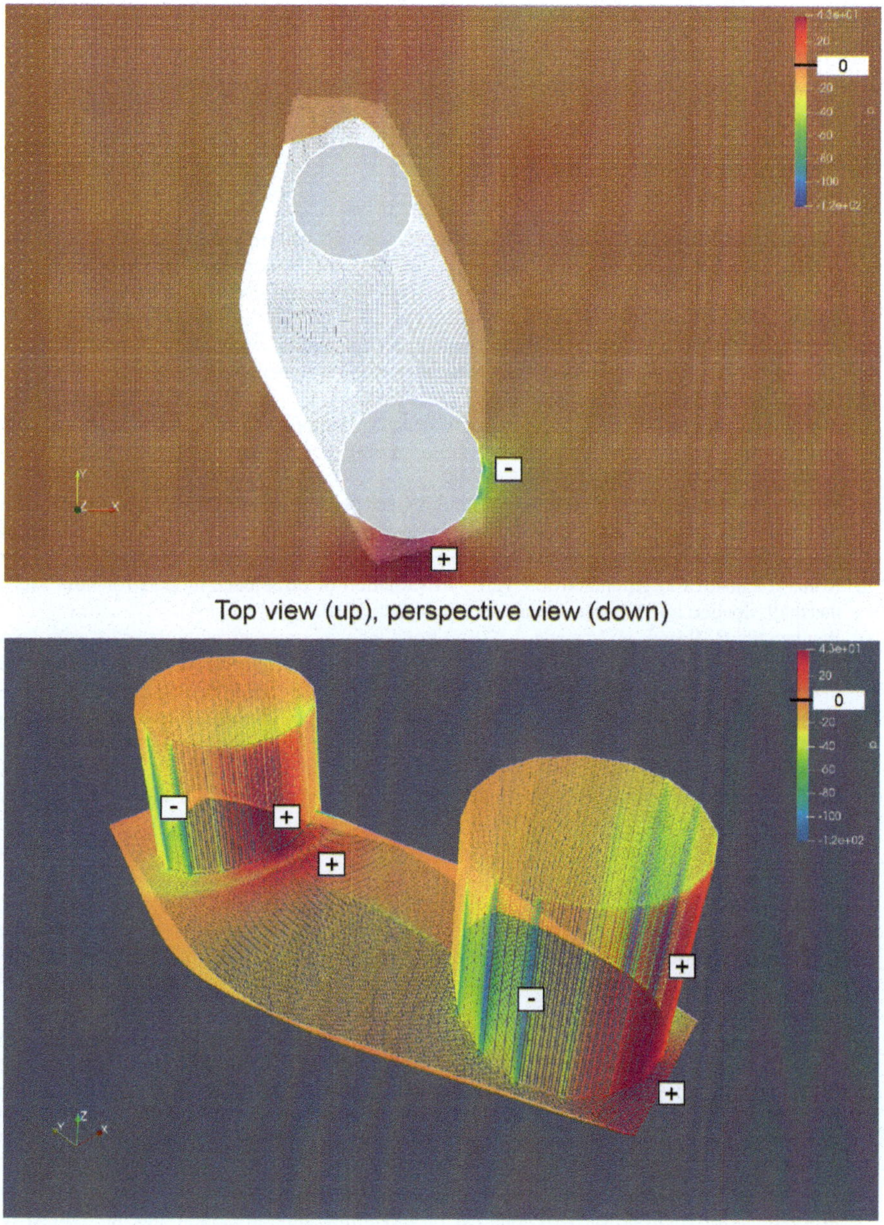


**Fig. 2.36** FlowBrane no. 3—the wind pressure in the southerly wind gusts. Areas of the positive (+) and negative (−) wind pressure

(iii) physical wind tunnel can be used to validate the simulation results of one (or two) best-performing alternatives.

## References

1. Lenzholzer S, Brown RD (2016) Post-positivist microclimatic urban design research: a review. Landsc Urban Plan 153:111–121. https://doi.org/10.1016/j.landurbplan.2016.05.008
2. Groat L, Wang D (2013) Simulation research. In: Architectural research methods. Wiley, New Jersey
3. Lobelia (2020) Past climate explorer. https://era5.lobelia.earth/. Accessed 5 May 2020
4. EnergyPlus (2021) Weather data. https://energyplus.net/weather. Accessed 8 May 2020
5. Greenshields CJ (2019) The OpenFOAM 7 user guide
6. Fröhlich J, von Terzi D (2008) Hybrid LES/RANS methods for the simulation of turbulent flows. Prog Aerosp Sci 44:349–377. https://doi.org/10.1016/j.paerosci.2008.05.001
7. White FM (2011) Fluid mechanics, 7th edn
8. Blocken B, Stathopoulos T, van Beeck JPAJ (2016) Pedestrian-level wind conditions around buildings: review of wind-tunnel and CFD techniques and their accuracy for wind comfort assessment. Build Environ 100:50–81. https://doi.org/10.1016/j.buildenv.2016.02.004
9. Kološ I, Lausová L, Michalcová V (2019) Evaluation of turbulence models for flow over a thermally loaded hill. J Numer Anal Ind Appl Math 13:10–19
10. Wennersten R, Brandt N, Larsson Å (2008) Loudden—a controversial harbour for petroleum products in Stockholm. In: Filho W, Brandt N, Krahn D, Wennersten R (eds) Conflict resolution in coastal zone management. Peter Lang Publishing, Bern, p meq.2008.08319dae.002
11. NEN (2006) Wind comfort and wind danger in the built environment: Dutch wind nuisance standard
12. Janssen WD, Blocken B, van Hooff T (2013) Pedestrian wind comfort around buildings: comparison of wind comfort criteria based on whole-flow field data for a complex case study. Build Environ 59:547–562. https://doi.org/10.1016/j.buildenv.2012.10.012
13. Du Y, Mak CM, Kwok K et al (2017) New criteria for assessing low wind environment at pedestrian level in Hong Kong. Build Environ 123:23–36. https://doi.org/10.1016/j.buildenv.2017.06.036
14. Janssen W, Blocken B, Van Hooff T (2013) Use of CFD simulations to improve the pedestrian wind comfort around a high-rise building in a complex urban area. In: 13th Conference of international building performance simulation association, Chambéry, France, August 26–28. Chambéry, France, pp 1918–1925
15. Zahid Iqbal QM, Chan ALS (2016) Pedestrian level wind environment assessment around group of high-rise cross-shaped buildings: effect of building shape, separation and orientation. Build Environ 101:45–63. https://doi.org/10.1016/j.buildenv.2016.02.015
16. Blocken B, Carmeliet J (2004) Pedestrian wind environment around buildings: literature review and practical examples. J Therm Envel Build Sci 28:107–159. https://doi.org/10.1177/1097196304044396
17. Welahettige P, Vaagsaether K (2016) Comparison of OpenFOAM and ANSYS fluent. In: Proceedings of EUROSIM 2016 congress on modelling and simulation and the 57th SIMS conference on simulation and modelling, Oulu, Finland, pp 1005–1012
18. Heisler GM, Dewalle DR (1988) 2. Effects of windbreak structure on wind flow. Agric Ecosyst Environ 22/23:41–69. https://doi.org/10.1016/0167-8809(88)90007-2
19. Dierickx W, Gabriels D, Cornelis WM (2002) Wind tunnel study on oblique windscreens. Biosyst Eng 82:87–95. https://doi.org/10.1006/bioe.2002.0048
20. Wu N, Wang Q, Xie X (2013) Wind energy harvesting with a piezoelectric harvester. Smart Mater Struct 22. https://doi.org/10.1088/0964-1726/22/9/095023

# Chapter 3
# Wind-Adaptive Building Envelope

This chapter contains three sub-chapters:

1. Virtual model
2. Physical model experiments
3. Building envelopes application

The wind-driven design leads us from influencing the airflow by an urban arrangement within the built environment, through the architectural form-finding process, integrating the wind fluxes, to finally affecting the wind on a smaller, building envelopes scale. Admittedly, it is not only the global building form or the urban configuration of buildings influencing the wind. A rough building surface, or small facade elements, can participate in the overall building aerodynamics and repress the unfavorable wind effects [1, 2].

A vision of adaptive architecture, morphing in real-time under the dynamic wind force, is investigated through a hybrid design method. First, one module of the wind-adaptive envelope is created digitally through Grasshopper and its plug-ins, such as Kangaroo 2, K2 Engineering, and Karamba 3D, for modeling the geometry, observing its response under the wind force in real-time, and consequently analyzing the structure, respectively. Digital designing is intertwined with physical prototyping to understand spatial relations and material constraints. The empirical material and geometry tests retroactively inform the settings needed in the simulations.

The motivation for choosing the tensegrity structural system as the wind-adaptive envelope was discussed in the section Tensegrity structures. A 4-strut tensegrity is coated with a lightweight tensile membrane substituting some of the elements in tension inside the tensegrity system. One wind-adaptive tensegrity-membrane module consists of nine 4-strut tensegrities. A network of wind-adaptive modules can create a double-layered building skin that articulates the wind that moves around it. The compressed components and bracing cables are placed in-between the outer and inner membrane (Fig. 3.1).



L. Kabošová et al., *Designing with the Wind*, Digital Innovations in Architecture, Engineering and Construction, https://doi.org/10.1007/978-3-031-24441-4_3

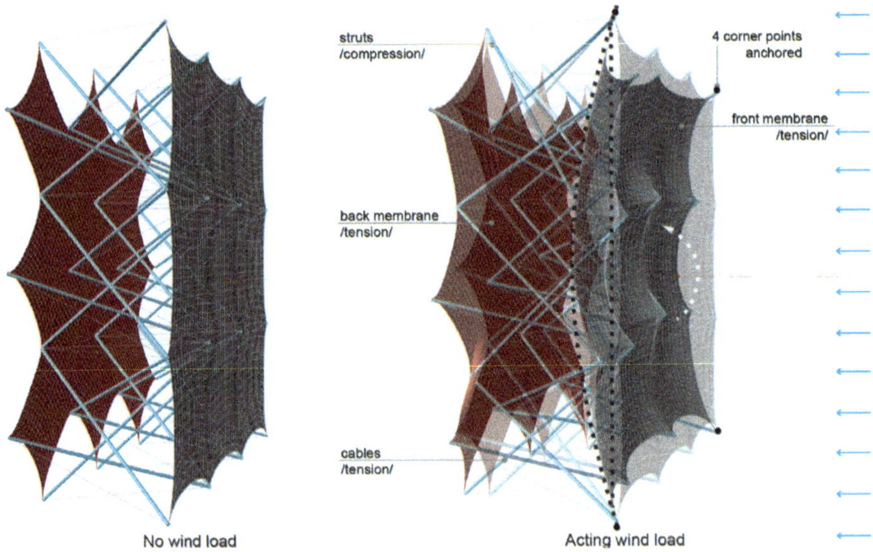


**Fig. 3.1** The wind-adaptive module, consisting of nine (3 x 3) 4-strut tensegrities with front and back tensile membrane. The bending is caused by the wind force, acting on the system

## 3.1 Virtual Model

One tensegrity-membrane module is analyzed in more detail. The elasticity of the tensegrity system under loads, combined with the material-embedded elasticity of the tensile fabric, allows the module, anchored in four front corner points, to bend under wind load (Fig. 3.1). This shape adaptation is reversible and contributes to the change in the overall aerodynamics of a building. No computer control or electrical driving mechanism is employed. The passive shape response is driven by the wind-geometry relations, enabling the real-time adaptation to the acting wind force.

### *3.1.1 Real-Time Response in the Wind*

Kangaroo 2 is employed to find the equilibrium state of the scaled tensegrity-membrane module considering the acting loads. The plugin utilizes dynamic relaxation, a vector-based method, which determines the static equilibrium form through the pseudo-dynamic equilibrium [3]. The digital and physical models of the structure are scaled 1:4. For adaptive-geometry studies, this scale adequately reproduces a full-scale prototype [4]. Settings for the digital simulations, realistically defining the material characteristics of the tensile fabric, wooden struts, and nylon cables, are explored simultaneously through digital and physical experiments. The material properties of the tensile fabric are analyzed through digital experiments with one 4-strut tensegrity. In Kangaroo 2, the pretension equal to 0 is maximal, whereas 1 means no pretension (Fig. 3.2). The following settings are selected as they correspond with the behavior of the fabric in reality: the stiffness k = 300, the pretension = 0.8. The material characteristics and the dimensions of wooden struts and nylon cables are depicted in Table 3.1.

In Kangaroo 2 simulations, the wind, acting on the structure, is represented by pressure in Pascal. The wind pressure on the surface is calculated from Bernoulli's equation with the value of 1.25 kg/m$^3$ used as air density. Within the computational model, the tensegrity-membrane module responds to the applied wind force in real-time. The settings from Table 3.1 are used to find the equilibrium state of the tensegrity-membrane module. After adding the wind load and the self-weight of wooden struts, the synergy of acting forces and material properties leads to the bending of the module.

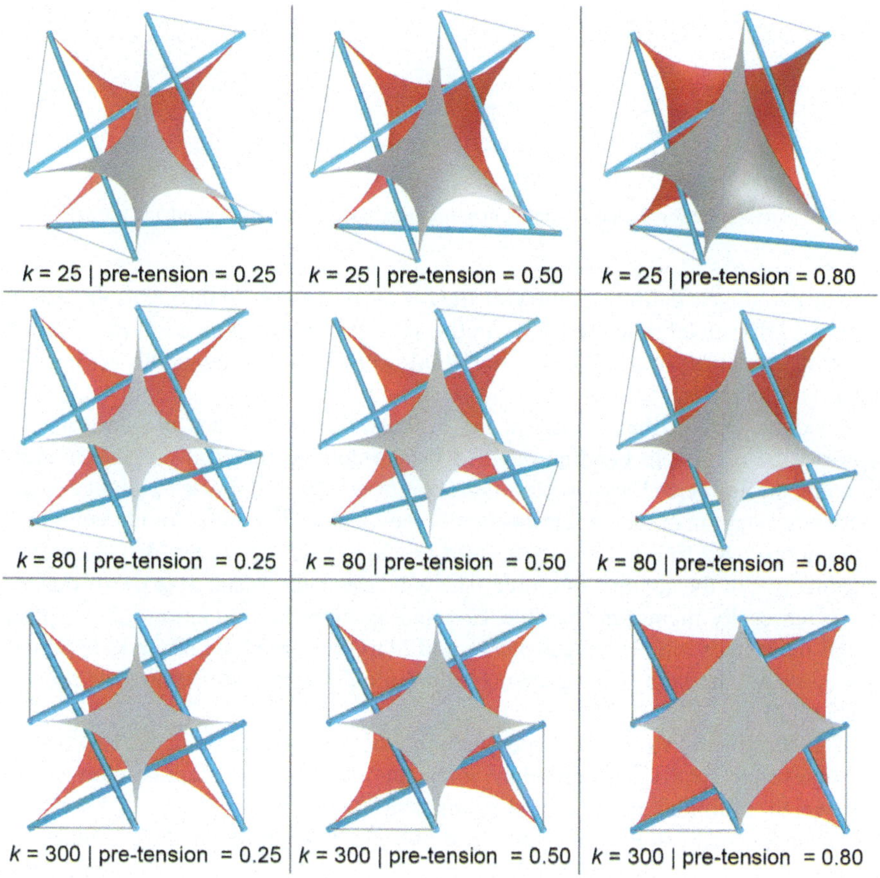


**Fig. 3.2** Kangaroo 2 experiments with one tensegrity-membrane unit

**Table 3.1** Material properties of the tensegrity-membrane module (scaled 1:4), defined in Kangaroo 2

| Type | Length/[m] | Diameter $\Phi$ or thickness $h$ [m] | Young's modulus of elasticity $E$ [Pa] | Stiffness $k$ [N/m] | Pretension (0 = maximum, 1 = no pretension) |
|---|---|---|---|---|---|
| Natural beechwood struts | 0.25 | $\Phi = 10e^{-3}$ | $46e^{6}$ in compression, in the direction of fibers | 14 444 | 1 |
| Nylon cables | 0.19 | $\Phi = 0.3e^{-3}$ | $3e^{9}$ | 1115 | 0.9 |
| Tensile fabric membranes | Front side 0.54 × 0.54 | $h = 0.5e^{-3}$ | $0.6e^{6}$ | 300 | 0.8 |

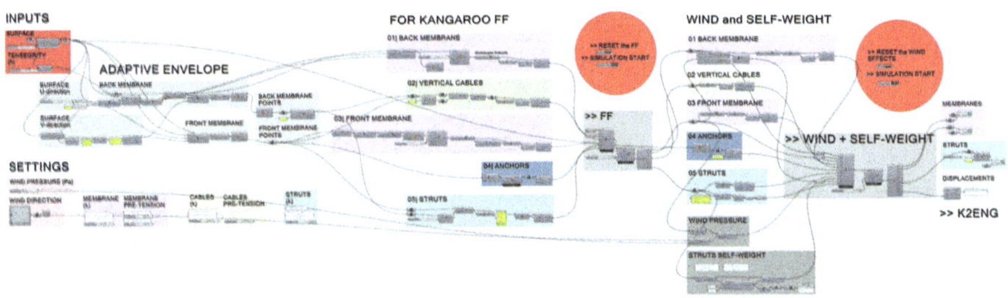


**Fig. 3.3** Finding the force equilibrium with subsequent loading with the wind and the self-weight

The parametric definition, showing the form-finding and, in the next step, loading the tensegrity-membrane module with the wind force, is displayed in Fig. 3.3. The resulting structure in the form-found equilibrium state (without the acting wind load) is depicted in Fig. 3.4.

Loading tests of the module in the 5, 9, and 24 m/s winds are performed to observe the reversible bending enabled by the defined material properties. Additionally, the counter-clockwise configuration of tensegrities causes a slight counter-clockwise rotation of the tensegrity-membrane module (Fig. 3.5). Another plugin for Grasshopper, K2 Engineering, is utilized to analyze the maximum displacement in the direction of the acting wind force. The wind acts perpendicular to the frontal side of the tensegrity-membrane module. Concluding from the simulations, the displacements induced by the wind speed of 5, 9, and 24 m/s are 56.34, 62.88, and 133.06 mm, respectively. The input wind pressure values are derived from Bernoulli's equation (Table 3.2).

The beechwood struts are loaded with the gravitational force in K2 Engineering to achieve physically correct behavior. In this scaled wind-adaptive element, the self-weight of thin membranes and nylon cables is negligible.

When the load 360 Pa, representing 24 m/s winds, is applied in the computational model, the struts penetrate through the membrane. That is not due to the calculation error. Albeit all elements in the simulation are linked, and their position is interdependent, the wind always acts on the mesh geometry only. The struts and cables are represented by springs connecting two points in space. The reason is the stretched

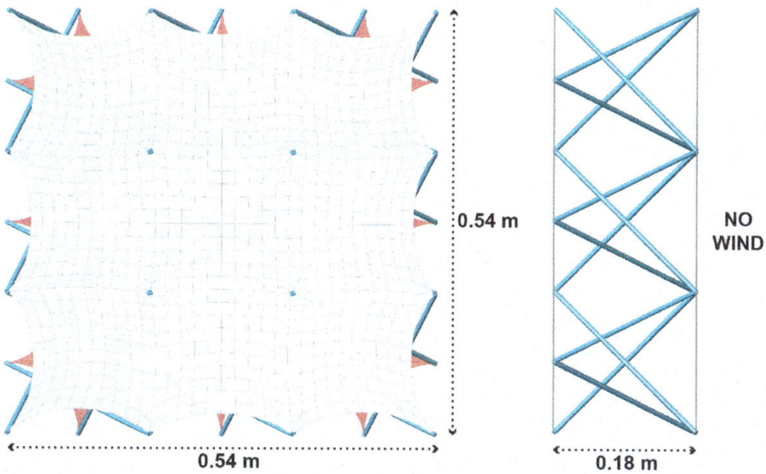


**Fig. 3.4** The proposed tensegrity-membrane module in the form-found state (left—front view, right—top view of the structure). The nylon bracing cables are not displayed

membrane geometry, which undergoes shape deformation in the acting force while the anchor points stay in place. As is confirmed by the physical experiment, this model set-up is correct.

## 3.2 Physical Model Experiments

The hybrid design strategy (digital experiments complemented with physical prototyping) is employed to create the tensegrity-membrane wind-adaptive module and analyze its behavior in the wind. Firstly, only one 4-strut tensegrity is assembled from beechwood struts and a rubber band (Fig. 3.6). Secondly, the tensegrity network is constructed by interconnecting nine units (3 x 3) into one module. The strut elements are attached using metal hooks, creating joints. The supporting panel with laser-cut holes serves for joint anchoring or pushing them in the upward direction. Slacked and taut cables are observed (Figs. 3.7 and 3.8). Although these cables are no longer in tension, the entire system of the tensegrity module redistributes itself to find an equilibrium of all acting forces.

The next step is substituting some cables with the tensile fabric (Figs. 3.8 and 3.9). An elastic knitted textile (90% micro-polyester and 10% elastane) in brown color is utilized for the first experiments. However, the (brown-colored) fabric is too stretchy, so it is over-stretched to hold the entire tensegrity-membrane network in a statically stable form. The fabric edges attached to the wooden struts are excessively strained (Fig. 3.9). Another two attempts with different materials resulted in a more balanced composition of artificial and natural fibers.

**Fig. 3.5** The front and top view of the tensegrity-membrane module subjected to different wind speeds. The nylon bracing cables are not displayed

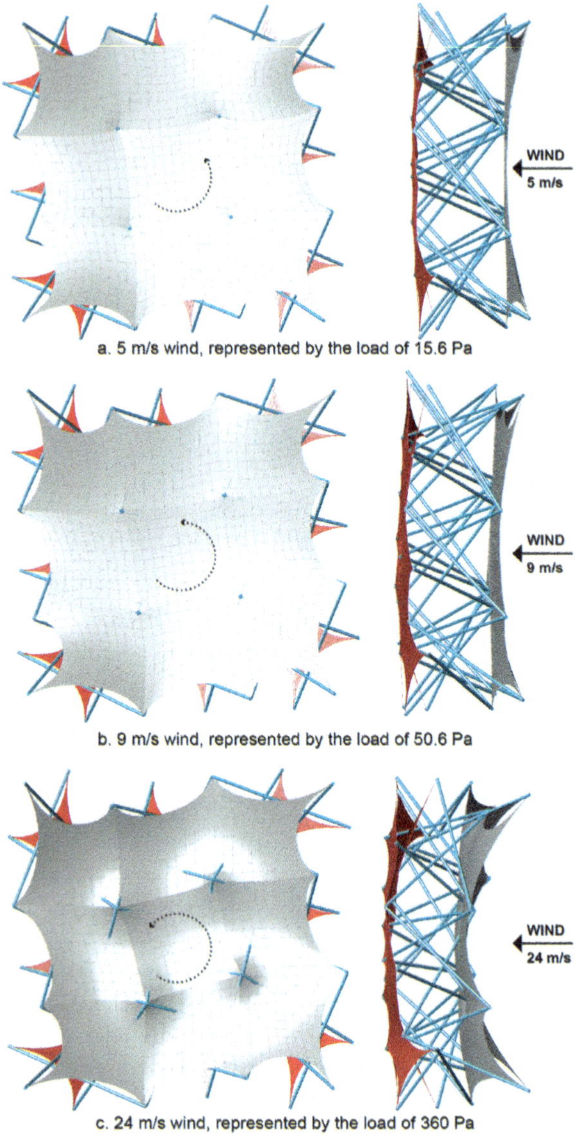


**Table 3.2** The displacement of the wind-adaptive module

| Wind speed | Wind pressure [Pa] | Displacement [mm] | Weight [kg] |
|---|---|---|---|
| 5 m/s | 15.6 | 56.34 | 0.61 |
| 9 m/s | 50.6 | 62.88 | 0.61 |
| 24 m/s | 360 | 133.06 | 0.61 |

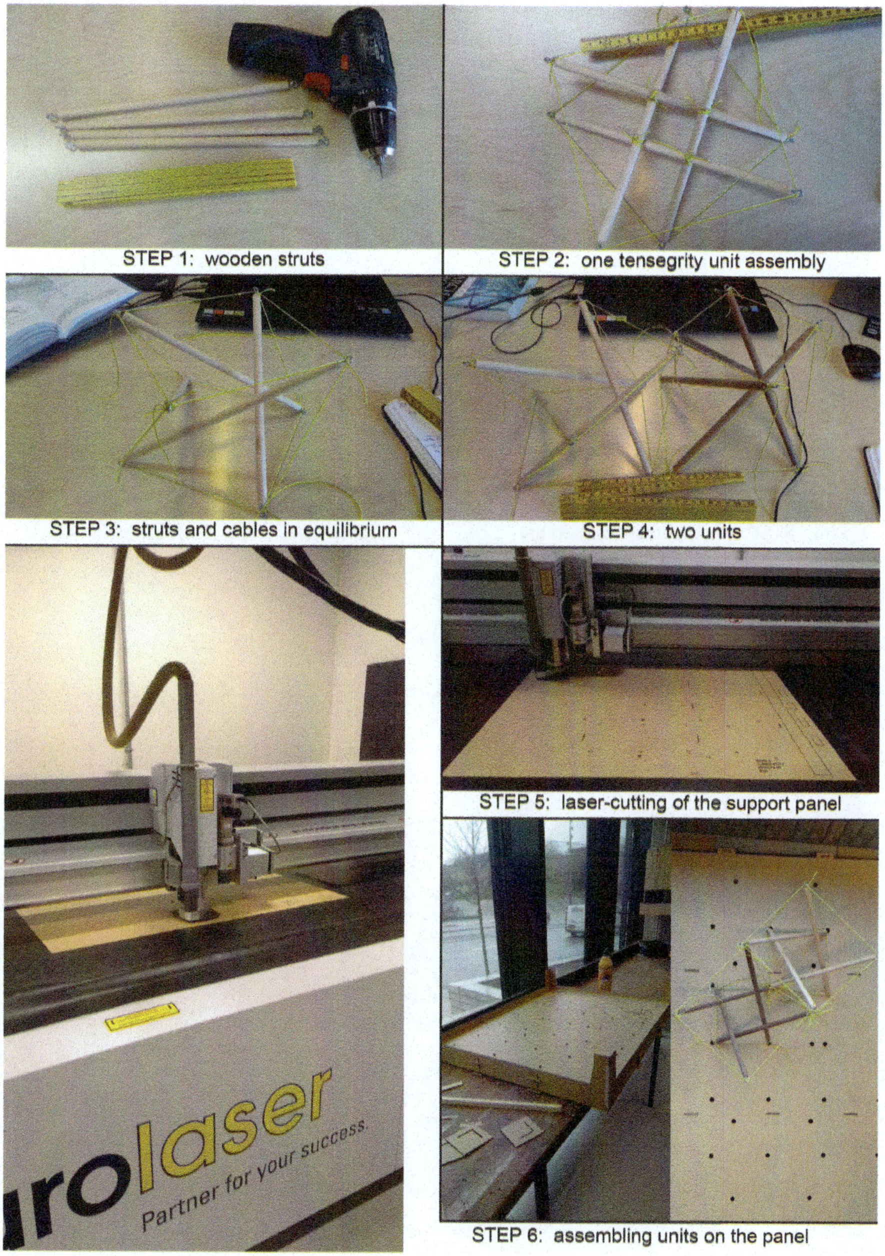


**Fig. 3.6** First experiments with the physical model

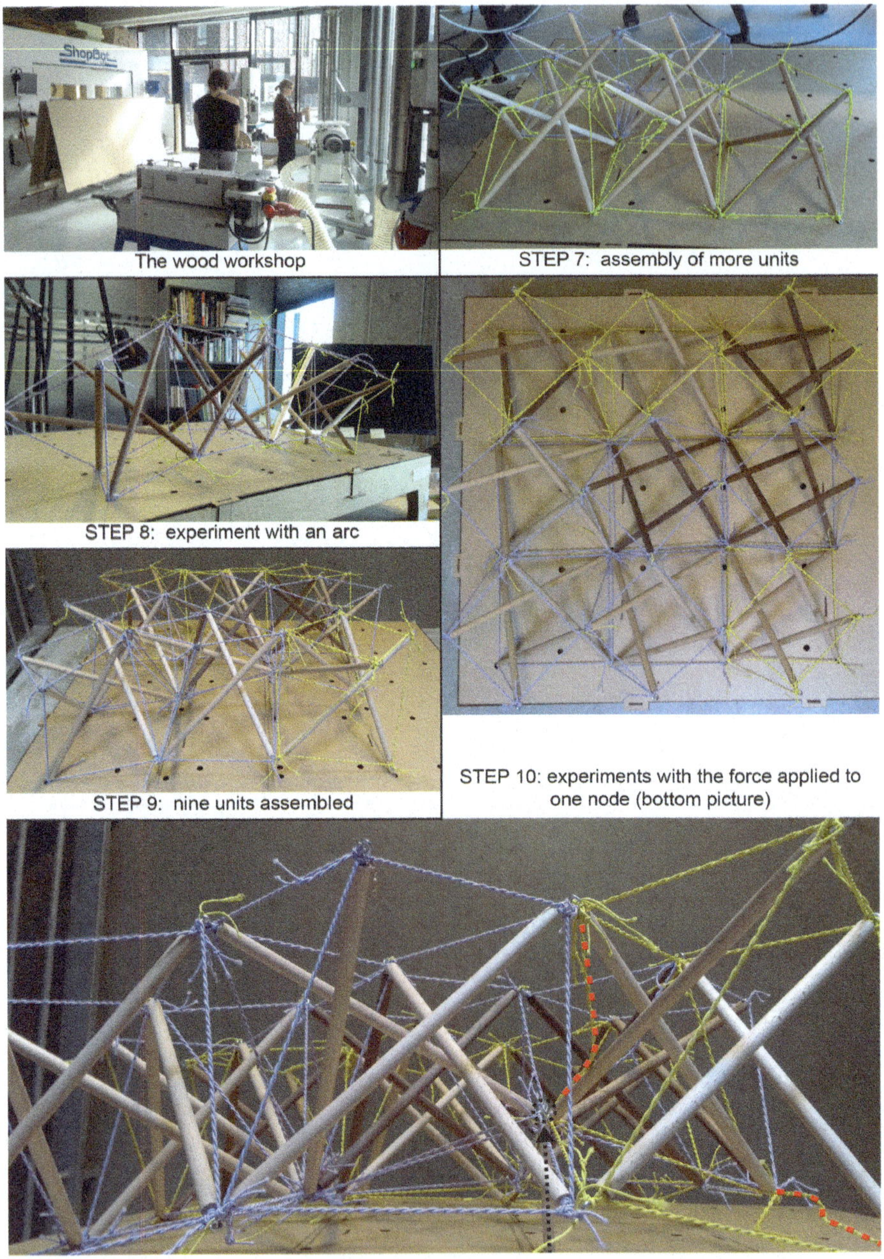


**Fig. 3.7** Physical experiment with the tensegrity network. The black dotted arrow represents the applied load. The red dotted curve represents the slack cables after the load is applied

**Fig. 3.8** Experiments with deformation of the tensegrity network. Experiments with fabrics

The yellow textile used for the final prototype is stiff yet elastic, consisting of 60% viscose, 35% polyamide, and 5% elastane (Fig. 3.9). In the digital simulations, the pretension of the fabric is 0.8. In the physical prototype, to create the same fabric pretension, the individual membrane pieces are laser-cut to 80% of their final size. That guarantees the stability of the structure. The individual parts of the fabric are sewn together with high tenacity, low-shrinkage thread to ensure a stiffer connection between each unit while forming a ribbed stiffening that can transpose the wind load

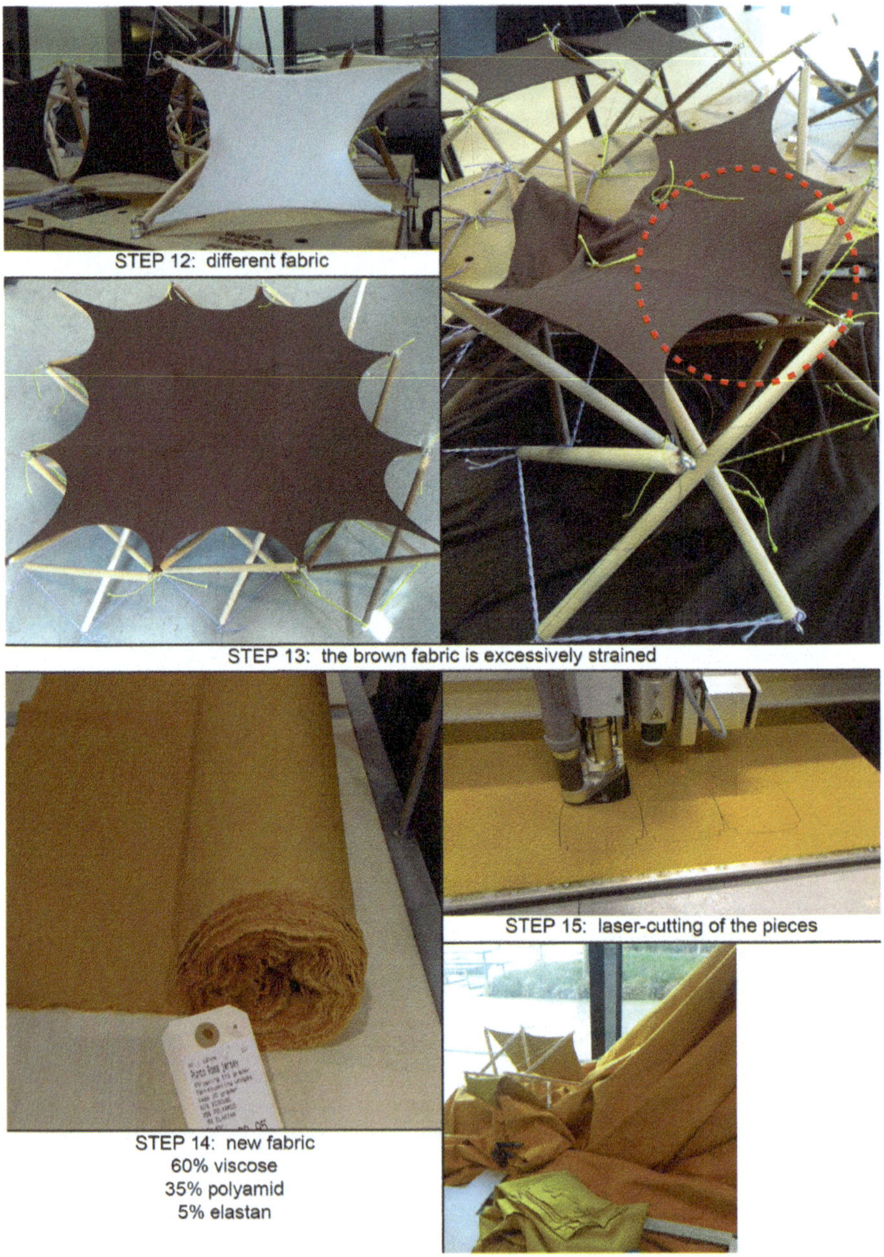


**Fig. 3.9** Experiments with various fabrics

from one node to another one (Fig. 3.10). The prototype is anchored in four front corners of the structure, exactly as it was with the digital model. It is then tested in the dynamic outdoor wind conditions by a fjord in Aalborg (Denmark).

The WeatherFlow measuring device (consisting of a thermometer, barometer, and anemometer) is utilized with the mobile phone application to obtain information about outdoor conditions during several 1-min measurements. The following data with the strongest wind gusts are documented here: temperature 13.2 °C, humidity 50%, average wind speed 4.6 m/s, and the maximum gusts 7.8 m/s.

The chaotic wind behavior implies that the outdoor experiments related to wind are challenging, as the wind gusts are random. Therefore the prototype is photographed and filmed. The wind-adaptive module is stabilized in outdoor conditions by a metal frame. Four corners of the front membrane are anchored to the frame while the structure hovers above the ground (Fig. 3.11). The joints, comprising four struts meeting at one point, are connected through metal hooks and attached to the fabric using a nylon thread. The metal frame anchors the prototype and stabilizes the structure in the wind. The metal ground-anchoring system is applied to the lower part of the frame to support the tensegrity-membrane module in the strong gusts (Fig. 3.12).

The prototype was installed on a small hill to subject the model to the prevailing westerly wind direction. The high grass might have somewhat interfered with the approaching wind flow, decelerating it near the ground.

Under the wind load, the designed tensegrity-membrane module bends and rotates counter-clockwise (when observed from the front side) in the same manner as in the computer simulations. Because the direction and speed of the wind flow are very dynamic and chaotic, the photo and film documentation focus on the maximum bending during wind gusts. The structure in the average-velocity wind, interlaced with instant wind gusts, can be observed in three video frames (Fig. 3.12).

The three overlapping pictures capture the bending of approximately 4 cm (Fig. 3.13). In the digital model, the bending in 5 m/s winds is around 5.6 cm, which confirms the similar behavior of the digital model and the prototype. The exact bending magnitude and rotation angle are directly related to the material characteristics, which, in the Kangaroo 2 simulations, are defined through the stiffness and pretension values. In the digital model, the joints are created by an ideal connection of all members, coinciding at one point, which is not achieved in the prototype. Even though the simulations are not 100% copies of reality, they can represent real-life situations [5]. The hybrid design approach helped validate the shape change of the wind-adaptive façade module in the wind, ergo the bending movement and rotation in the counter-clockwise direction. To conclude, the Kangaroo 2 simulations can reliably predict the dynamic response of the designed structure in the wind.

The following section will deal with the digital modeling of a full-scale adaptive module in Kangaroo 2 and subsequent structural analysis in Karamba 3D. Networking multiple tensegrity-membranes together creates a coherent, watertight wind-adaptive building envelope. Each tensegrity-membrane module could be anchored to a lightweight steel gridshell, creating a load-bearing support frame (Fig. 3.14).

**Fig. 3.10** Physical experiment—sewing and final steps

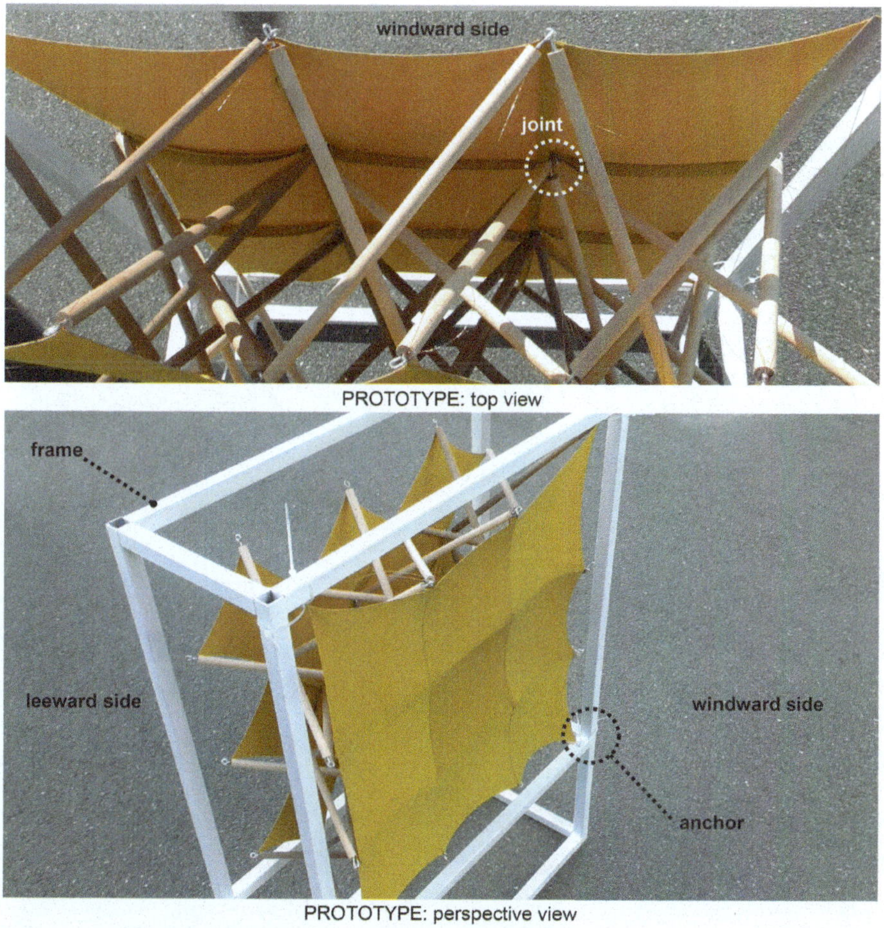


**Fig. 3.11** The system of the tensegrity-membrane prototype

**Fig. 3.12** The side view of the prototype. From left to right: no wind, average speed, and wind gust

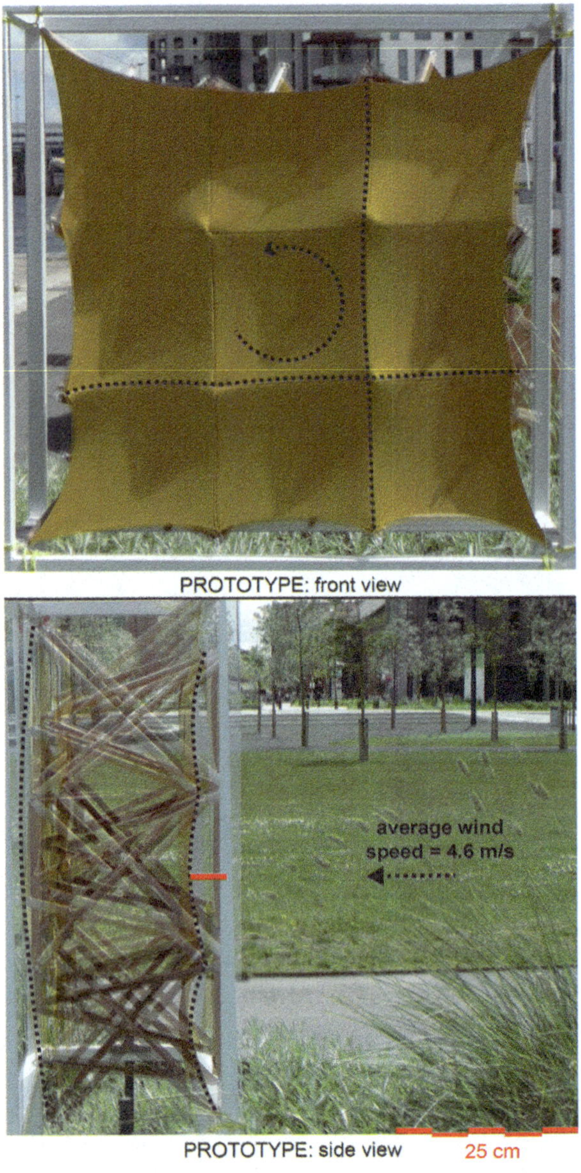


**Fig. 3.13** The wind tests of the physical prototype

## 3.3 Building Envelopes Application

The hybrid examination of the scaled tensegrity-membrane model is followed by the digital investigations of a full-scale module (3 x 3 units). This lightweight structure weighs 9.1 kg and comprises beechwood struts, stainless steel cables, and two layers of structural tensile membranes (Fig. 3.14).

**Fig. 3.14** The lightweight structure of the wind-adaptive building envelope

Compared to the scaled model, the material attributes of struts, cables, and fabric are different, resulting in less distinctive bending of the structure in the wind. Nevertheless, when the full-scale wind-adaptive module is applied to a building skin, a potentially new reading of wind through architecture is achieved from both the exterior and interior (Fig. 3.15).

**Fig. 3.15** Wind-adaptive tensegrity-membrane building envelope

### 3.3.1 *The Full-Scale Wind-Adaptive Module*

For the digital assembly and loading of the full-scale wind-adaptive module, Kangaroo 2 is complemented with K2 Engineering, as in the case of the scaled module. The axial stresses in membranes and cables are determined through K2 Engineering. The values of axial stresses will be necessary as initial strain values for further structural analysis in Karamba 3D. It is decided to use wooden struts with a thinner cross-section and analyze whether the designed dimensions will suffice. The dimensions of the components, as well as their material characteristics, are displayed in Table 3.3.

With the settings in Kangaroo 2 from Table 3.3, the lightweight, full-scale adaptive module bends slightly less, and the rotating movement is less evident (Fig. 3.16).

The behavior of the scaled structure and the full-scale module is compared only digitally in Kangaroo 2. The relation of the speed of the acting wind to the magnitude of maximum displacement is pictured in Fig. 3.17. The continuous line depicts the bending of the scaled model, while the dashed line depicts the bending of the full-scale module in the wind. The displacement is similar in the higher wind speeds.

Karamba 3D, a Finite Element Method (FEM) parametric structural engineering plugin for Grasshopper, is employed to verify that the designed dimensions of the wind-adaptive module could work in reality. The material features of the beechwood struts, stainless steel cables, and tensile membrane are depicted in Table 3.4. Karamba 3D does not work with the basic SI unit system, so the material properties are in Karamba-used units. The unit problem is sometimes confusing and is a minor drawback of this analysis tool.

Two load cases are investigated in Karamba 3D: the stress distribution in the form-found module and, sequentially, the stress distribution in the module under the wind load of 360 Pa, corresponding to the value of the 24 m/s wind speed. The values of the initial strain $\varepsilon_0$, defining the pretension of membranes and cables, are obtained from K2 Engineering using the following formula:

**Table 3.3** Material properties of the full-scale tensegrity-membrane element, defined in Kangaroo 2

| Type | Length/[m] | Diameter $\Phi$ or thickness $h$ [m] | Young's Modulus of elasticity $E$ [Pa] | Stiffness $k$ [N/m] | Pretension (0 = maximum, 1 = no pretension) |
|---|---|---|---|---|---|
| Natural beechwood struts | 1 | $\Phi = 16e^{-3}$ | $10.5e^9$ in compression, in the direction of fibers | 2 110 080 | 1 |
| Stainless steel cables | 0.78 | $\Phi = 2e^{-3}$ | $130e^9$ | 523 333 | 0.99 |
| Tensile fabric membranes | Front side 2.16 × 2.16 | $h = 0.51e^{-3}$ | $16.5e^7$ | 84 150 | 0.9 |

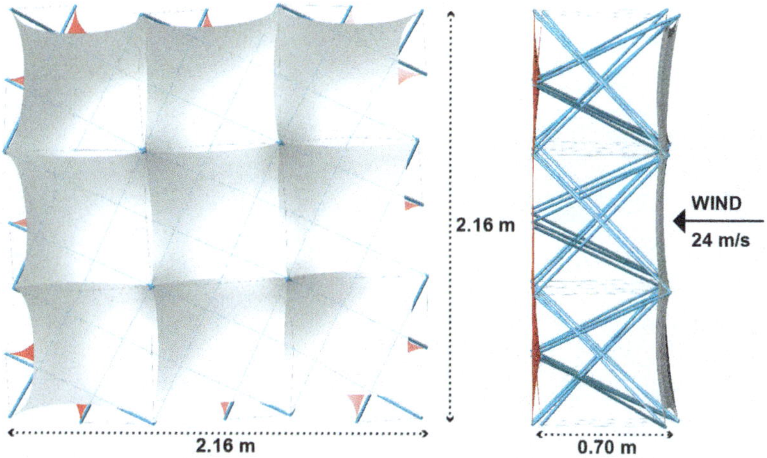


**Fig. 3.16** The bending of the full-scale wind-adaptive module in 24 m/s wind

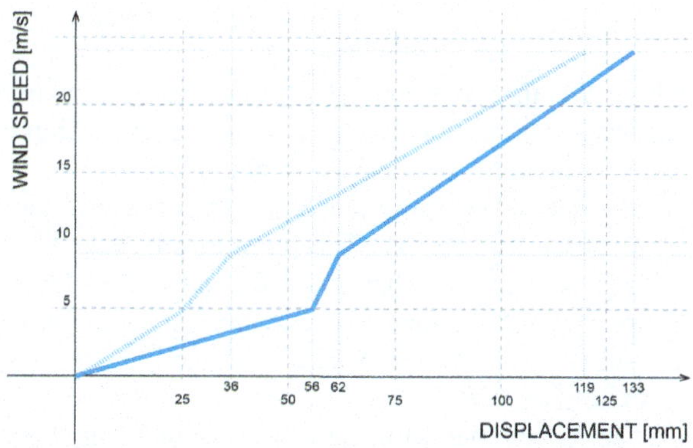


**Fig. 3.17** Bending in the wind of the scaled model (continuous line), in contrast to full-scale module (dashed line)

**Table 3.4** Material settings in Karamba 3D

| | Diameter $\Phi$ or thickness $t$ [cm] (material) | Specific weight $\gamma$ [kN/m$^3$] | Young's modulus $E_1$, $E_2$ [kN/cm$^2$] | In-plane shear modulus $G_{12}$ [kN/cm$^2$] | Transverse shear modulus $G_{31}$, $G_{32}$ [kN/cm$^2$] | Yield strength $fy_1$, $fy_2$ [kN/cm$^2$] |
|---|---|---|---|---|---|---|
| Struts | $\Phi = 1.6$ (wood) | 6 | 1050 | 360 | 360 | 2.5 |
| Cables | $\Phi = 0.2$ (steel) | 78.5 | 13 000 | 5000 | 5000 | 23.5 |
| Membranes (orthotropic) | $h = 0.051$ (fabric) | 5.7 | $E_1 = E_2 =$ 16.5 | 6.157 | 6.157 | 4.575 |

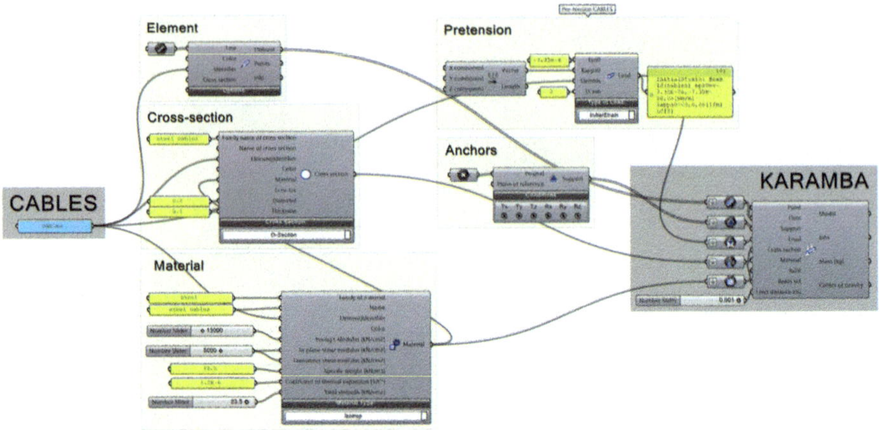


**Fig. 3.18** The Grasshopper definition of the cable elements within the tensegrity-membrane module

$$N = \varepsilon_0 A E \tag{3.1}$$

where $N$ is the axial force [kN], $\varepsilon_0$ is the initial axial strain imposed on the element (positive sign = elongation), $A$ is the cross-section of the element [cm$^2$], and $E$ is Young's modulus of elasticity [kN/cm$^2$].

The definition of the tensegrity-membrane model in Karamba 3D shall closely approximate the reality. The equilibrium state of the full-scale module is found through Kangaroo 2, after which the individual elements are defined in Karamba 3D. Cross-sections and material characteristics are linked to geometry. We subsequently add the pretension, represented by the initial strain $\varepsilon_{0,}$ and fix selected anchor points in space (Fig. 3.18).

The membranes within the pre-stressed adaptive module in the equilibrium state (Fig. 3.19) are subjected to Von Mises Stress of $+1.16\text{e}^{-5}$ to $+2.73\text{e}^{-3}$ kN/cm$^2$ (i.e., $+11.6\text{e}^{-5}$ to $+27.3\text{e}^{-3}$ MPa). The maximum compression in struts equals $-6.66\text{e}^{-4}$ kN/cm$^2$ (i.e., $-66.6\text{e}^{-4}$ MPa). The maximum tension in cables reaches $+2.21\text{e}^{-2}$ kN/cm$^2$ (i.e., +0.221 MPa).

The wind velocity of 24 m/s is used as the fundamental value of the basic wind velocity $v_{b0}$ in Stockho, Sweden (the location of the case study from the previous chapter). This wind speed value is applied in the simulations to analyze the tensegrity-membrane module under the wind load.

With all materials defined in Karamba 3D, the observed bending of the tensegrity-membrane module in the 24 m/s wind causes the displacement of 7.26 cm. Compared to the simulation in Kangaroo 2 (Fig. 3.16), the structure with exact, detailed material characteristics defined in Karamba 3D is stiffer and bends less under the same wind load (Fig. 3.20). The material properties of membranes are crucial because the membranes not only serve for capturing the wind but also for the force distribution to the rest of the structure.

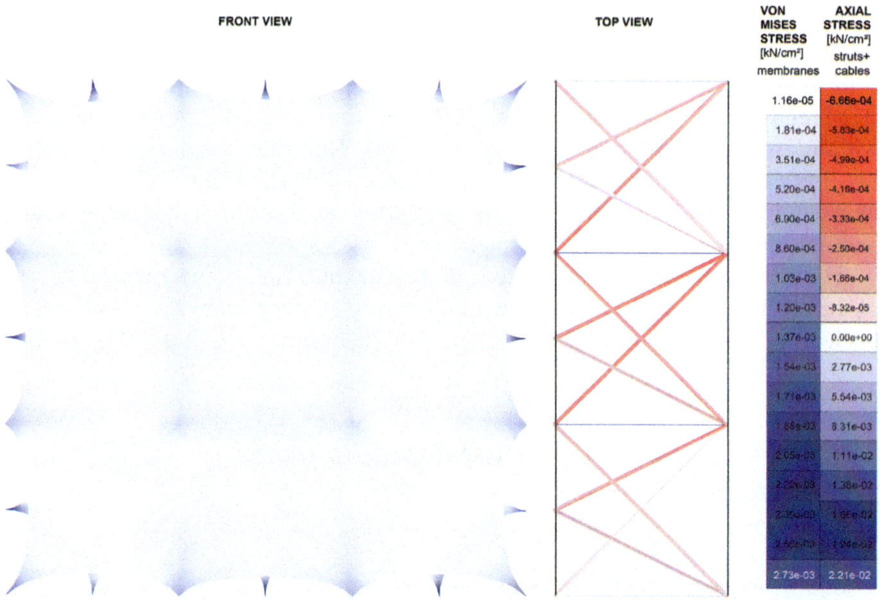


**Fig. 3.19** The wind-adaptive module in the form-found state. No wind load is applied. The tension is represented in blue, and the compression in red color

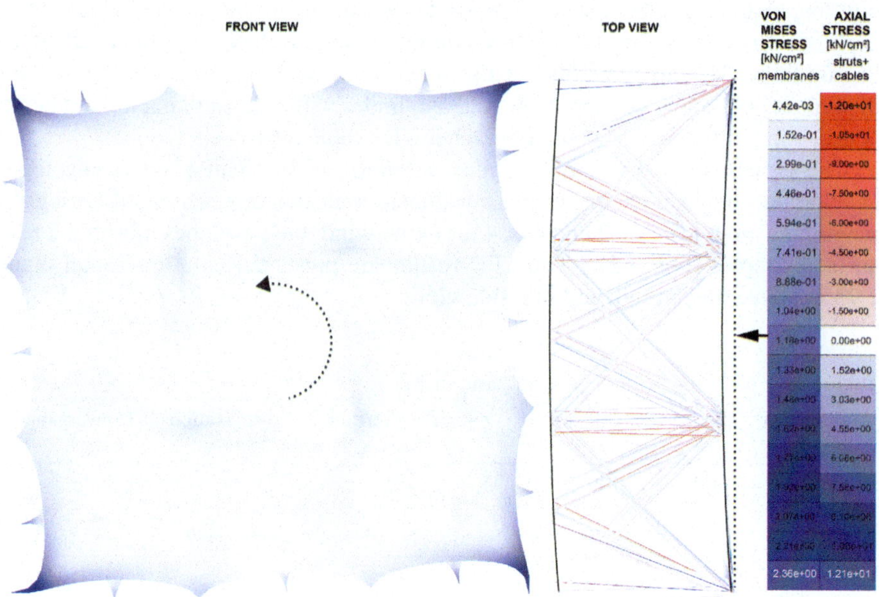


**Fig. 3.20** The wind-adaptive module under the load of 360 Pa, representing 24 m/s wind

The extreme value of the wind velocity induces the maximum von Mises Stress in membranes equal to +2.36 kN/cm$^2$ (i.e., +23.6 MPa) (Fig. 3.20). This value, compared to the yield strength from Table 3.4, implies that the membrane can withstand high wind speeds without damage. The maximum compression in struts is also within limits, −1.64 kN/cm$^2$ (i.e., −16.4 MPa). This value cannot be observed in Fig. 3.20. The maximum tension in cables equals +1.21e$^1$ kN/cm$^2$ (i.e., +121 MPa), which is again lower than the yield strength (the upper force limit that can be applied without causing a permanent deformation).

The passive, flexible adaptation enabled by the material and geometry properties of the elements ensures that the structure recovers to its original state once it is not loaded.

### 3.3.2 *Wind Analysis of Basic-Shaped Buildings with the Adaptive Envelope*

The proposed wind-adaptive module creates a dimpled effect on the building envelope under the acting wind load. Examples from other industries suggest that surface dimples can contribute to the surface pressure and even drag force reduction, in addition to altering the wind flow pattern [6, 7]. Experiments with basic-geometry building shapes, a cube, and a cylinder, tested in the 12 and 24 m/s winds, were performed to specify how the designed wind-adaptive envelope alters the wind effects on buildings. By employing Swift CFD, a cube with a 12 m side is tested in perpendicular and oblique wind approach, and a cylinder with a 12 m diameter and height is tested in perpendicular wind approach angle (Table 3.5).

The smooth-surface variants are juxtaposed to variants with adaptive envelopes. In addition, the adaptive envelope is evaluated with two dimensions of tensegrity-membrane modules, 2 × 2 and 4 × 4 m. In the wind, the envelopes morph to create surface dimples (Fig. 3.21). The CFD results are presented on a horizontal plane, 1.75 m above the ground, and in a side view.

**Table 3.5** VWT settings for the tests of basic shapes

| Cell size [m] | xy-scale | xy-offset | z-scale | Terrain roughness | Relaxation factors | Convergence criteria |
|---|---|---|---|---|---|---|
| 1 | 1.2 | 50 | 5 | 1 | $p = 0.3$<br>$v_{x,y,z} = 0.7$<br>$k\|\varepsilon = 0.7$ | $p = 4e^{-2}$<br>$v_{x,y,z} = 1e^{-2}$<br>$k\|\varepsilon = 1e^{-2}$ |

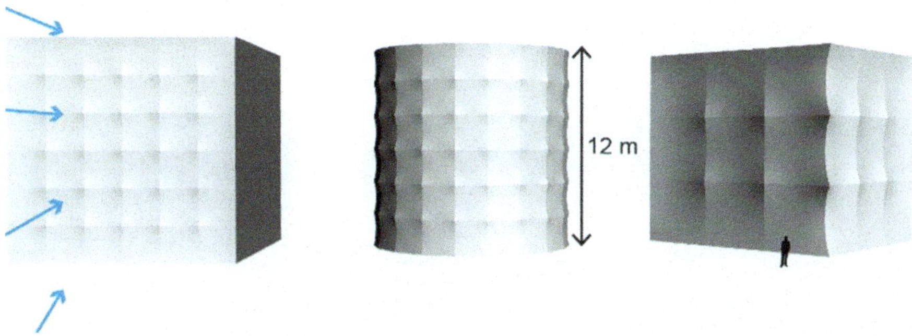


**Fig. 3.21** A cube, a cylinder, and an inclined cube with 2 × 2 or 4 × 4 m dimpled envelope

#### 3.3.2.1 The Cube in the Perpendicular Wind Approach

Only the windward side of the tested cube is coated in the wind-adaptive envelope. First, the cube is tested in 12 m/s winds. The bending of the membrane corresponds with the magnitude of the acting wind force, creating dimples. Through the post-processing in Paraview, the surface pressure, drag force, and drag coefficient are determined (Table 3.6).

From Table 3.6, a reduction in wind surface pressure is observed, indicating a slight decrease in wind suction when the adaptive envelope is applied. Albeit the dimpled surface is spiky-shaped, which should cause more drag in the wind, the drag coefficient is a little lower for the variant with 2 m dimples compared to the smooth variant. The simulations in the 12 m/s wind show no striking difference in the leeward flow pattern (Fig. 3.22).

The pressure distribution around the tested variants looks similar, too (Fig. 3.23). Hence, applying the adaptive envelope on the windward side of a cube seems not particularly favorable in the 12 m/s wind.

**Table 3.6** The cube in the 12 m/s wind

| | Surface pressure $p_k$ [m$^2$/s$^2$] | Drag force $F_d$ [N] | Frontal area [m$^2$] | Drag coefficient $c_d$ [–] | Iterations for CFD convergence | Surface suction dimpled versus smooth |
|---|---|---|---|---|---|---|
| Smooth cube | −96.16 to + 75.95 | −10 875 | 144 | 0.839 | 215 | – |
| Cube 2 m dimples | −93.68 to + 75.96 | −10 807 | 144 | 0.834 | 215 | ↓2.6% |
| Cube 4 m dimples | −93.11 to + 75.99 | −10 987 | 144 | 0.848 | 213 | ↓3.2% |

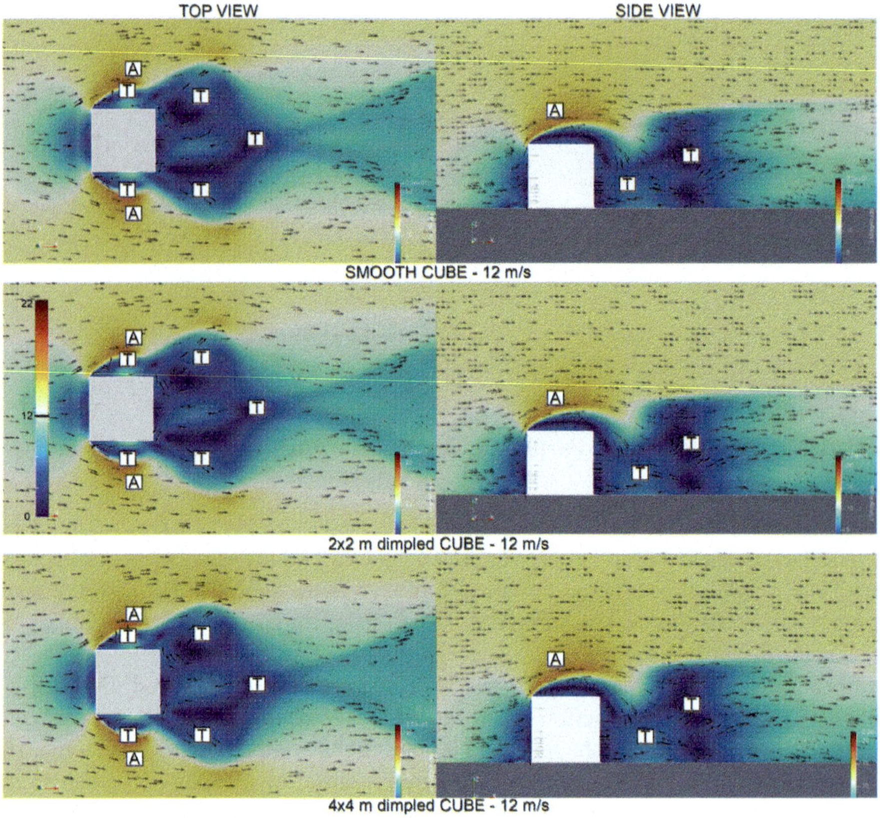


**Fig. 3.22** 12 m/s winds. The wind flow pattern and the wind speed of the three tested variants of the cube

The same variables are observed in double wind speed. They are displayed in Table 3.7.

The bending of the membrane, caused by the wind force, corresponds to wind magnitude. There is no significant reduction in the wind surface pressure in 24 m/s winds with the adaptive envelope applied on the windward side of the cube. Also, as expected, the drag coefficient rises when the dimpled surface is created.

The pattern of the wind flow, on the other hand, is being affected by the discernible spikes of the adaptive envelope placed on the windward side of the cube (Fig. 3.24). With the dimple-shaped envelope, the symmetrical leeward turbulent zones form closer to the tested object because the wind flow separates differently on the cube edges with the applied adaptive element.

Due to the different wind flow separation on the cube with the adaptive envelope, the suction on the lateral sides of the cube emerges closer to the front side of the examined object. Moreover, wind swirls occur right behind the cube and are very distinctive in the 4 × 4 m dimpled variant (Fig. 3.25).

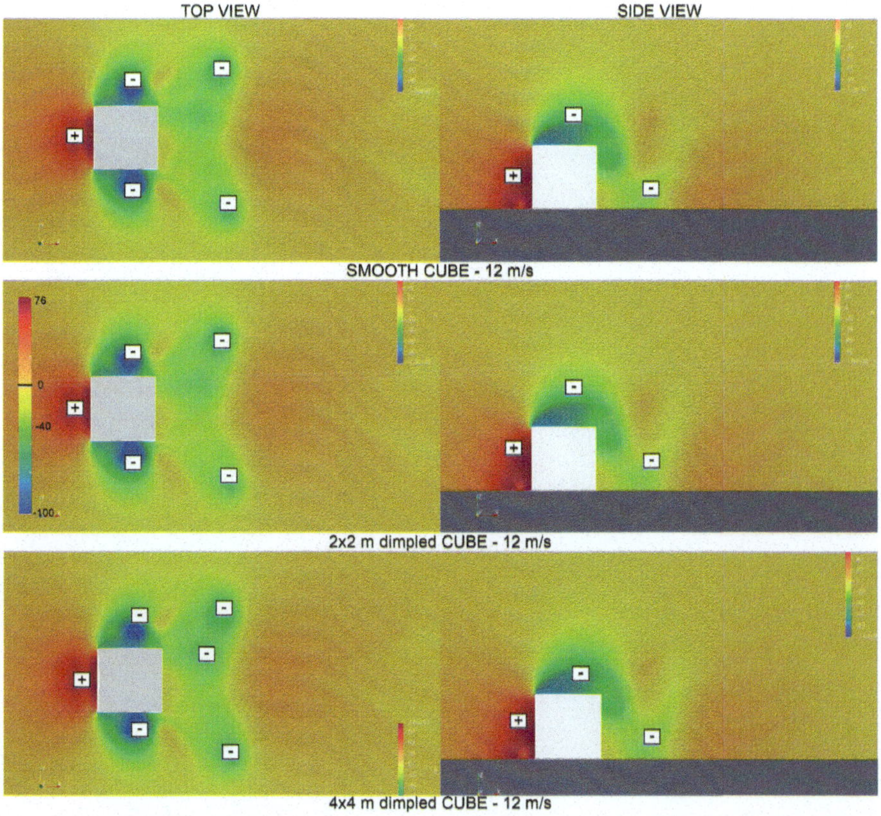


**Fig. 3.23** 12 m/s winds. The wind pressure distribution around the three tested variants of the cube

**Table 3.7** The cube in the 24 m/s wind

| | Surface pressure $p_k$ [$m^2/s^2$] | Drag force $F_d$ [N] | Frontal area [$m^2$] | Drag coefficient $c_d$ [–] | Iterations for CFD convergence | Surface suction dimpled versus smooth |
|---|---|---|---|---|---|---|
| Smooth cube | −375.37 to +302.41 | −44 088 | 144 | 0.85 | 229 | – |
| Cube 2 m dimples | −378.28 to +303.2 | −44 614 | 144 | 0.861 | 215 | ↑0.7% |
| Cube 4 m dimples | −368.49 to +303.9 | −45 351 | 144 | 0.87 | 206 | ↓1.8% |

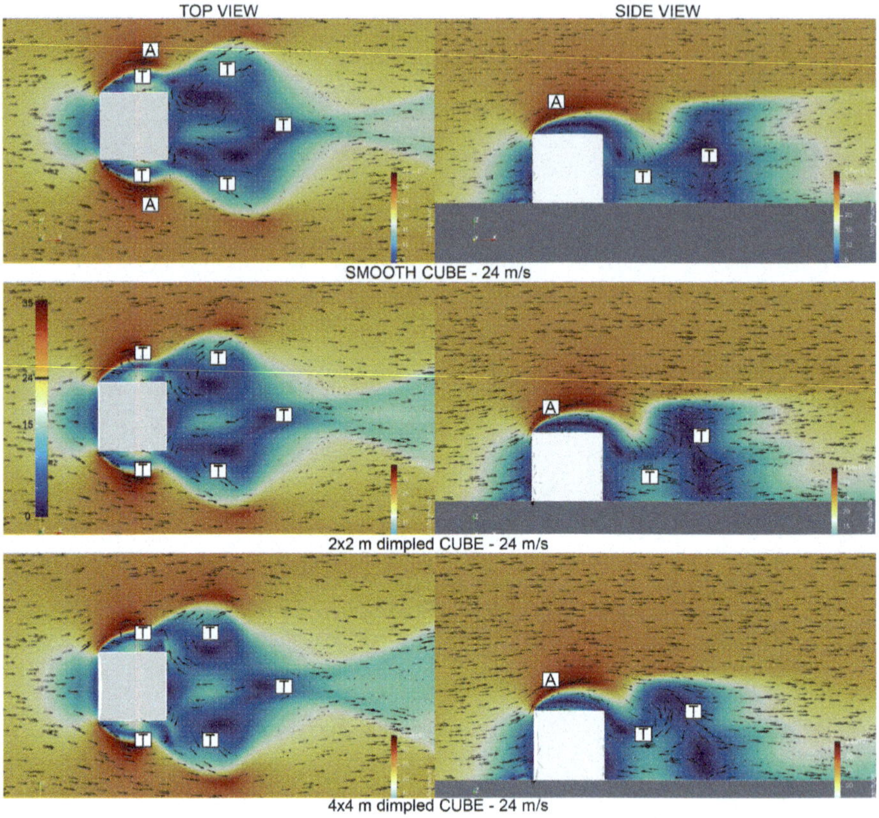


**Fig. 3.24** 24 m/s winds. The wind flow pattern and the wind speed of the three tested variants of the cube

#### 3.3.2.2 The Cylinder

The CFD simulation of a cylindrically-shaped building in 12 m/s winds suggests the following: despite the wind suction increases by 15 and 30.3% for the 2 × 2 m and 4 × 4 m dimpled variant, respectively, the aerodynamics of the cylinder is improved when the 2 × 2 m adaptive modules are applied on the whole envelope, leading towards the reduction of drag force, and hence drag coefficient (Table 3.8). However, with the 4 × 4 m adaptive elements, a drastic increase in the drag force and the drag coefficient occurs.

The influence of the wind-adaptive envelope on the wind flow pattern is very distinct in the case of the cylindrical building (Fig. 3.26). A calm zone emerges on the leeward side.

Moreover, the dimpled surface reduces the zones of wind suction on the lateral sides of the cylinder (Fig. 3.27).

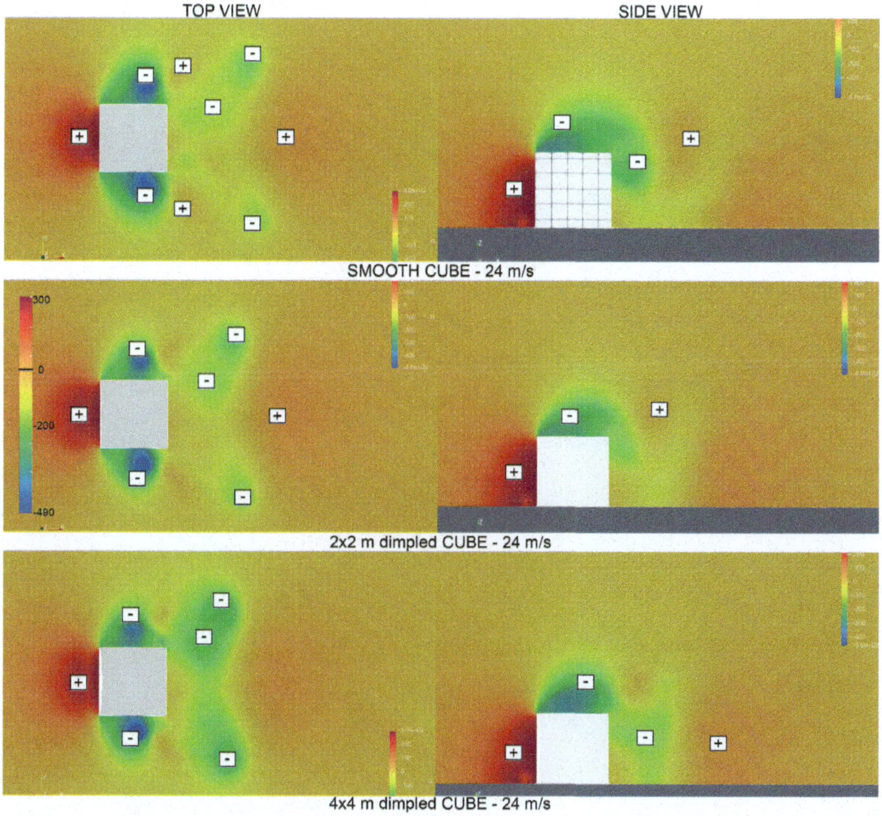


**Fig. 3.25** 24 m/s winds. The wind pressure distribution around the three tested variants of the cube

**Table 3.8** Cylinder $\Phi = 12$ m, $d = 12$ m in the 12 m/s wind

| | Surface pressure $p_k$ [m$^2$/s$^2$] | Drag force $F_d$ [N] | Frontal area [m$^2$] | Drag coefficient $c_d$ [–] | Iterations for CFD convergence | Surface suction dimpled versus smooth |
|---|---|---|---|---|---|---|
| Smooth cylinder | −180.14 to +73.93 | −3 834.2 | 144 | 0.296 | 604 | – |
| Cylinder 2 m dimples | −207.25 to +73.92 | −3 586.74 | 144 | 0.277 | 295 | ↑15% |
| Cylinder 4 m dimples | −234.71 to +73.94 | −16 029 | 144 | 1.24 | 360 | ↑30.3% |

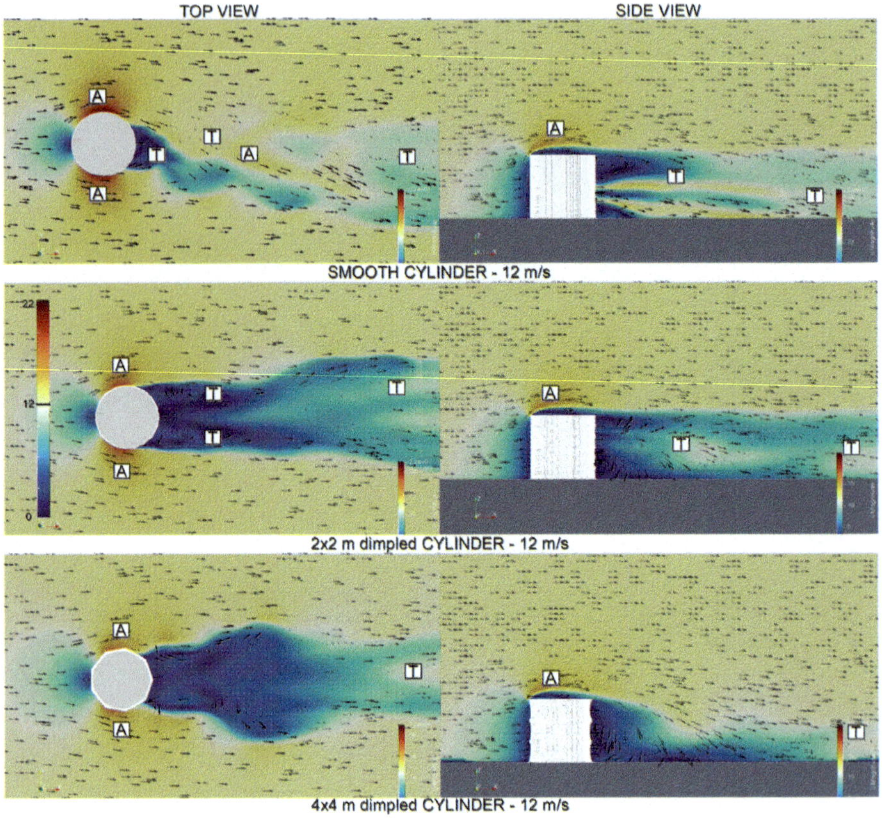


**Fig. 3.26** 12 m/s winds. The wind flow pattern and the wind speed of the three tested variants of the cylinder

Albeit the wind surface suction increases with the application of the wind-adaptive envelope, the drag coefficient of the cylindrical building is reduced, and a calmer downwind zone is created through the 2 × 2 adaptive modules applied on the cylinder's surface.

In the twofold (i.e., 24 m/s) wind speed, the wind suction increases for the wind-adaptive envelope variants. However, in stronger winds, too, the 2 × 2 variant contributes to reducing the drag force and the drag coefficient of the cylinder (Table 3.9).

Like the CFD results in the 12 m/s wind, calm symmetrical wind swirls are created but disappear on the pedestrian level (in the case of the 4 × 4 m dimpled surface) (Fig. 3.28).

Although generally, it might seem that the surface suction is not reduced with the employed wind-adaptive envelope, the leeward side suction and the strong suction on the lateral sides of the cylinder can be positively affected (Fig. 3.29).

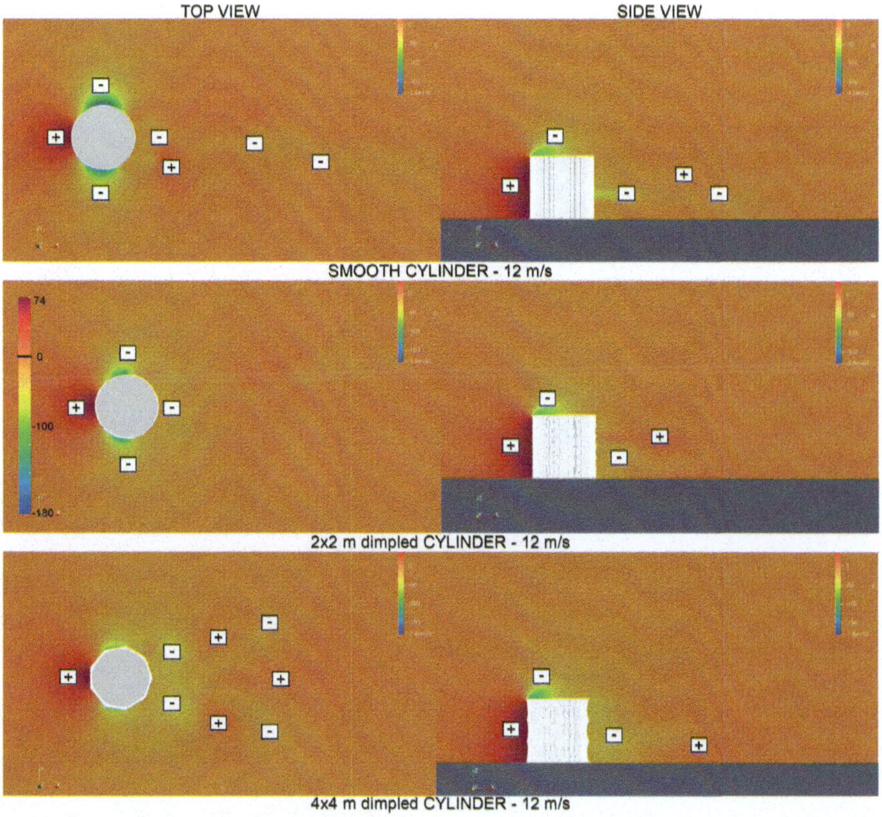


**Fig. 3.27** 12 m/s winds. The wind pressure distribution around the three tested variants of the cylinder

**Table 3.9** Cylinder $\Phi = 12$ m, $d = 12$ m in the 24 m/s wind

| | Surface pressure $p_k$ [m$^2$/s$^2$] | Drag force $F_d$ [N] | Frontal area [m$^2$] | Drag coefficient $c_d$ [–] | Iterations for CFD convergence | Surface suction dimpled versus smooth |
|---|---|---|---|---|---|---|
| Smooth cylinder | −748.23 to +296.36 | −16 317 | 144 | 0.315 | 542 | – |
| Cylinder 2 m dimples | −815.41 to +295.81 | −14 979 | 144 | 0.289 | 326 | ↑9% |
| Cylinder 4 m dimples | −1005.24 to +296.43 | −21 362 | 144 | 0.412 | 340 | ↑34.3% |

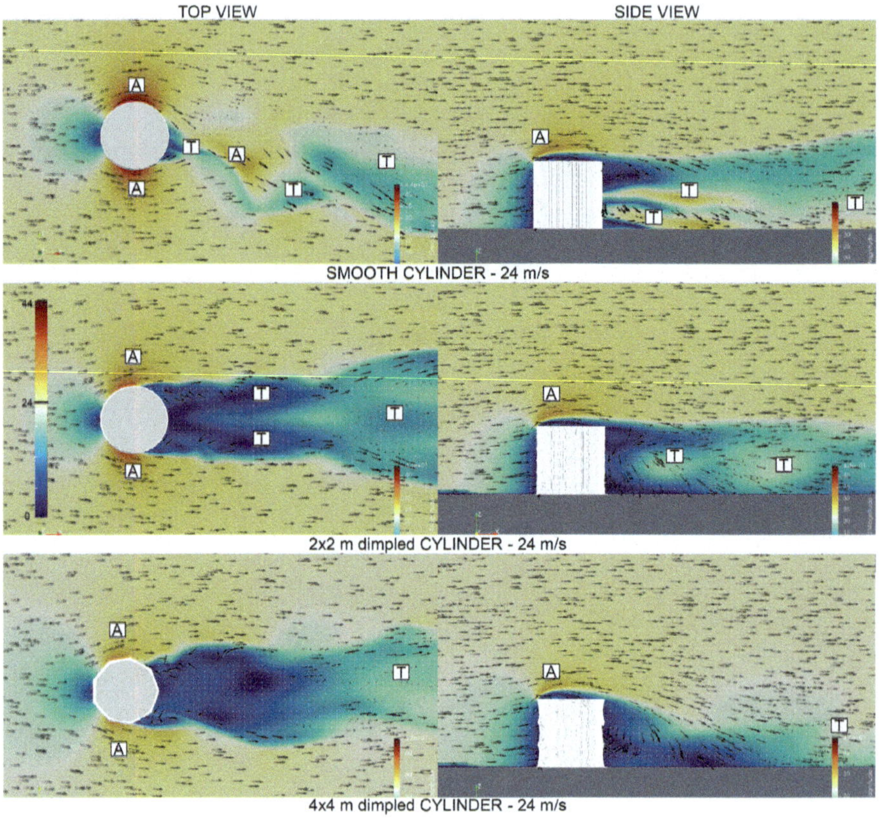


**Fig. 3.28** 24 m/s winds. The wind flow pattern and the wind speed of the three tested variants of the cylinder

#### 3.3.2.3 The Cube in the Oblique Wind Approach

If the developed wind-adaptive envelope is applied on two sides of the cube, or, in other words, if the wind approach is oblique to the cube, the wind surface pressure slightly increases for both the 2 × 2 m and 4 × 4 m variant. The drag force and drag coefficient are almost unchanged (Table 3.10).

The side edges of the inclined cube induce wind flow separation. The flow, except for the zones of symmetrical turbulent swirls, is calmer at the pedestrian level when the wind-adaptive envelope is applied (Fig. 3.30). The suction zones are similar for all three variants (Fig. 3.31).

Doubling the wind speed does not cause the drag force or the drag coefficient to rise in the case of a 2 × 2 m variant, contrary to the 4 × 4 m variant (Table 3.11). The wind surface suction, on the other hand, increases by 30.1% when the 2 × 2 m dimples are applied, compared to the smooth cube variant. The wind surface suction is again reduced when the adaptive skin with 4 × 4 m modules is employed.

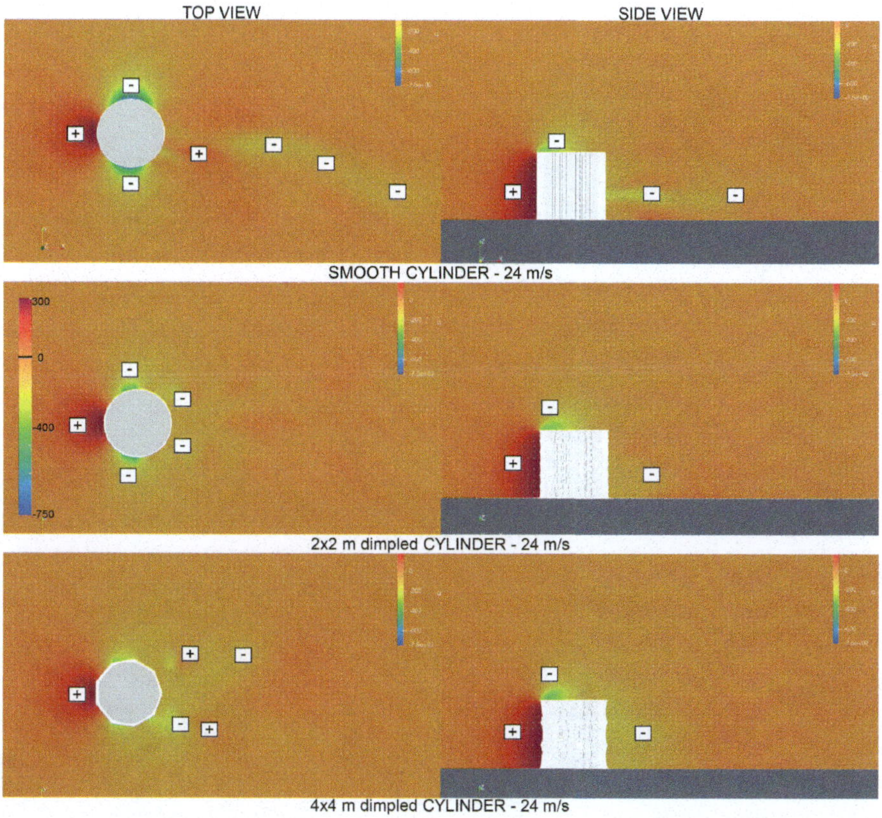


**Fig. 3.29** 24 m/s winds. The wind pressure distribution around the three tested variants of the cylinder

**Table 3.10** 45°-inclined cube in the 12 m/s wind

| | Surface pressure $p_k$ [$m^2/s^2$] | Drag force $F_d$ [N] | Frontal area [$m^2$] | Drag coefficient $c_d$ [–] | Iterations for CFD convergence | Surface suction dimpled versus smooth |
|---|---|---|---|---|---|---|
| Smooth rotated cube | −226.59 to +73.79 | −14 237 | 204 | 0.775 | 353 | – |
| Inclined cube 2 m dimples | −237.27 to +75.61 | −14 238 | 204 | 0.775 | 356 | ↑4.7% |
| Inclined cube 4 m dimples | −250.79 to +75.5 | −14 497 | 204 | 0.790 | 358 | ↑10.7% |

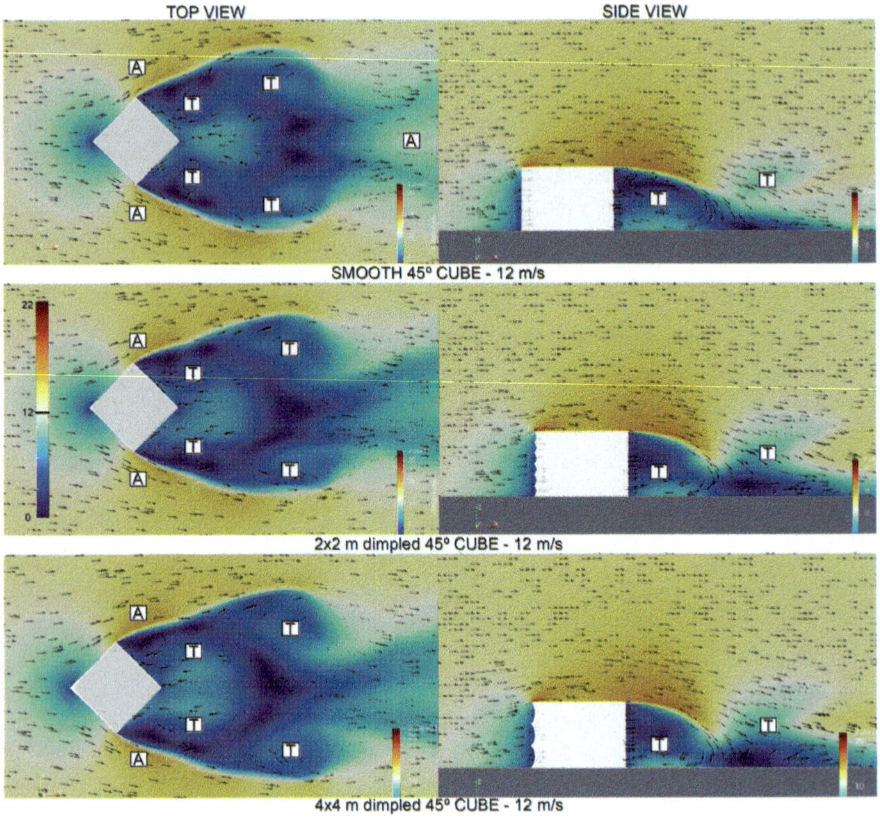


**Fig. 3.30** 12 m/s winds. The wind flow pattern and the wind speed of the three tested variants of the 45°-inclined cube

The symmetrical leeward wind swirls at the pedestrian level are more visible when the wind-adaptive envelope is engaged. However, the wind speed on the leeward side is decreased compared to the smooth shape variant. The dimple effect on the façade generates a calmer downwind zone. That is even more distinctive in the alternative with 2 × 2 m dimples (Fig. 3.32).

The wind-adaptive envelope slightly reduces and dissipates the wind suction zones on the leeward side of the geometry (Fig. 3.33).

### *3.3.3 FlowBrane No. 3 with the Wind-Adaptive Envelope*

As the CFD tests of basic building shapes suggest, the efficiency and the effect of the designed wind-adaptive envelope depend on the global building form. To corroborate the assumption, FlowBrane no. 3, representing a streamlined building

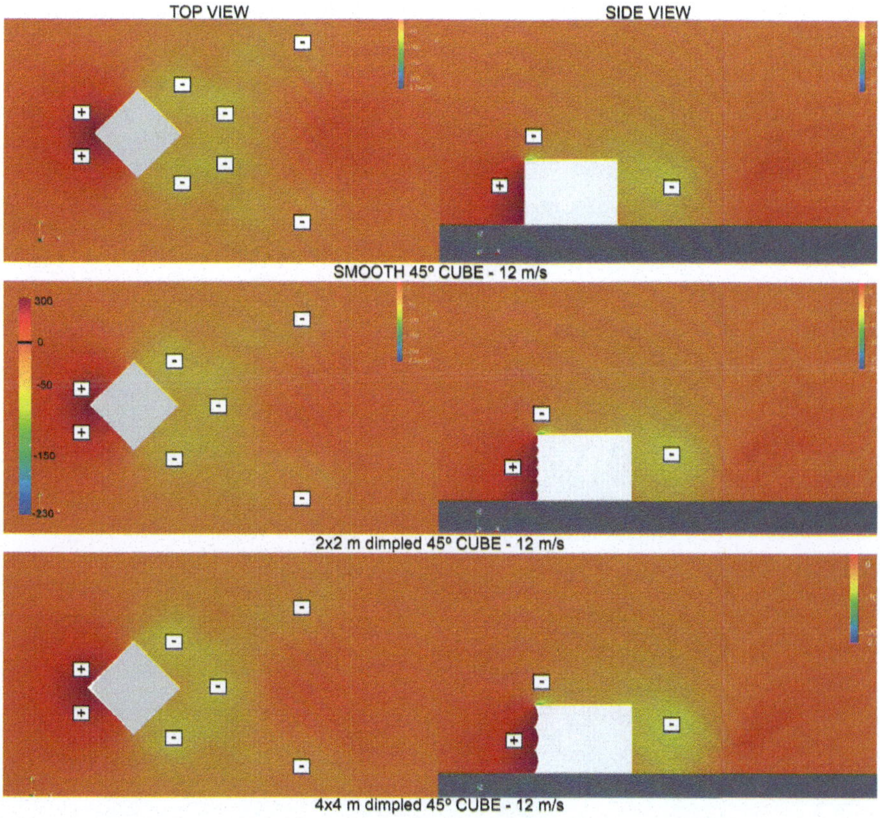


**Fig. 3.31** 12 m/s winds. The wind pressure distribution around the three tested variants of the 45°-inclined cube

**Table 3.11** 45°-inclined cube in the 24 m/s wind

| | Surface pressure $p_k$ [m$^2$/s$^2$] | Drag force $F_d$ [N] | Frontal area [m$^2$] | Drag coefficient $c_d$ [–] | Iterations for CFD convergence | Surface suction dimpled versus smooth |
|---|---|---|---|---|---|---|
| Smooth rotated cube | −908.75 to +294.24 | −57 785 | 204 | 0.787 | 345 | – |
| Inclined cube 2 m dimples | −1182.39 to +298.20 | −57 666 | 204 | 0.785 | 335 | ↑30.1% |
| Inclined cube 4 m dimples | −1033.61 to +299.0 | −60 130 | 204 | 0.819 | 335 | ↑13.7% |

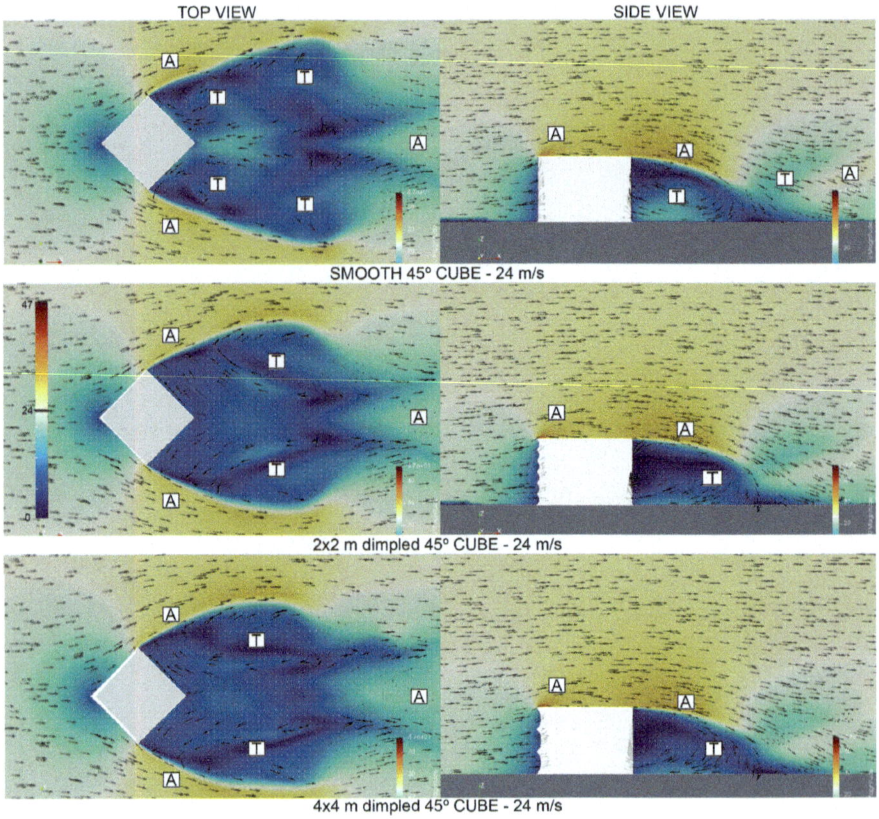


**Fig. 3.32** 24 m/s winds. The wind flow pattern and the wind speed of the three tested variants of the 45°-inclined cube

geometry, is tested in 24 m/s wind in these versions: a smooth architectural shape wrapping the two cylindrical silos and an adaptive envelope instead of the smooth one. The CFD simulations of the cylindrically-shaped building demonstrated that the size of the adaptive module of 2 × 2 m decreases the drag force and the drag coefficient. Although the wind surface suction increases by 9% in the 24 m/s wind, an experiment is carried out. The proposed wind-adaptive envelope with the size of the adaptive module of 2 × 2 m is used not only on the shape of FlowBrane no. 3 but also on the two adjacent cylinders (silos). The settings of the CFD simulations are the same as during the previous testing of FlowBrane no. 3 (in the section CFD wind analysis through Swift for Grasshopper). The simulation results are depicted in Fig. 3.34.

The wind-adaptive building envelope reduces the leeward side turbulence occurring right behind the silo of the first wind impact. The situation downwind from the FlowBrane no. 3 is calmer with the adaptive envelope. Indeed, it is the wind pressure that, acting on the tensegrity-membrane building skin, ultimately reduces the wind

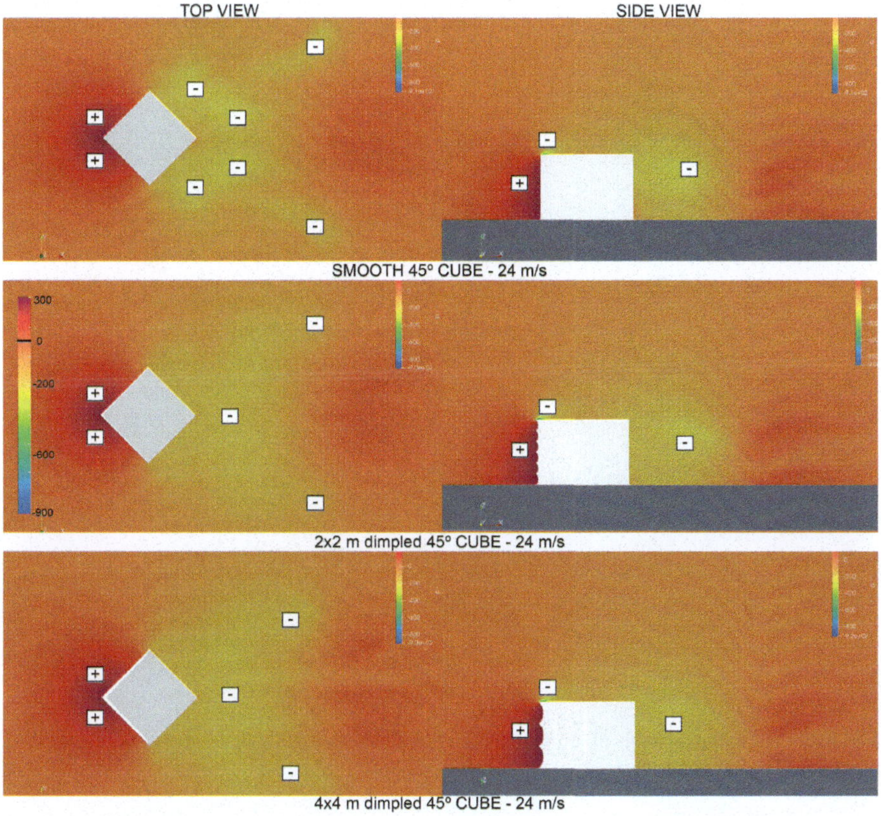


**Fig. 3.33** 24 m/s winds. The wind pressure distribution around the three tested variants of the 45°-inclined cube

suction induced by FlowBrane no. 3. The wind suction (negative value) is reduced by 28.98%, while the wind pressure (positive value) by 2.7% in the 24 m/s wind (Table 3.12).

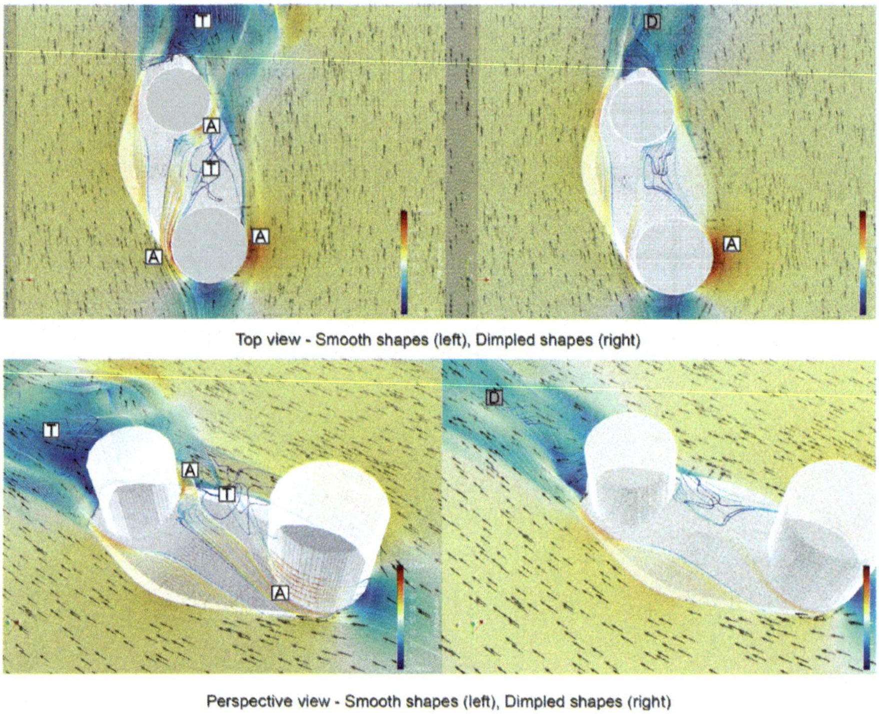


**Fig. 3.34** The wind analysis of smooth and dimpled shapes in 24 m/s wind. (A) accelerated, (T) turbulent, (D) diffused flow

**Table 3.12** FlowBrane no.3 in the 24 m/s wind

| | Surface pressure $p_k$ [$m^2/s^2$] | Drag force $F_d$ [N] | Frontal area [$m^2$] | Drag coefficient $c_d$ [–] | Iterations for CFD convergence | Surface suction dimpled versus smooth |
|---|---|---|---|---|---|---|
| Smooth FlowBrane no. 3 | −929.26 to +297.79 | −5 923.68 | 404.14 | 0.041 | 219 | – |
| FlowBrane no. 3 with 2 m dimples | −659.93 to +289.77 | −6 452.78 | 404.14 | 0.044 | 278 | ↓28.98% |

# References

1. Lignarolo L, Lelieveld C, Teuffel P (2011) Shape morphing wind-responsive facade systems realized with smart materials. In: Proceedings of the adaptive architecture conference. London, UK, pp 1–14
2. Giacchetti A, Bartoli G, Mannini C (2019) Wind effects on permeable tall building envelopes: issues and potentialities. CTBUH J. 20–27

3. Barnes MR (1988) Form-finding and analysis of prestressed nets and membranes. Comput Struct 30:685–695. https://doi.org/10.1016/0045-7949(88)90304-5
4. Veuve N, Sychterz AC, Smith IFC (2017) Adaptive control of a deployable tensegrity structure. Eng Struct 152:14–23. https://doi.org/10.1016/j.engstruct.2017.08.062
5. Groat L, Wang D (2013) Simulation research. In: Architectural research methods. Wiley, Inc., New Jersey, USA
6. Chear CK, Dol SS (2015) Vehicle aerodynamics: drag reduction by surface dimples. Int J Mech Mechatronics Eng 9:202–205
7. Al-Obaidi KM, Azzam Ismail M, Hussein H, Abdul Rahman AM (2017) Biomimetic building skins: an adaptive approach. Renew Sustain Energy Rev 79:1472–1491. https://doi.org/10.1016/j.rser.2017.05.028

# Chapter 4
# Conclusions

As a reaction to the era of climate change, the book proposes a wind-driven design method as an innovative alternative to established design. It strives to contribute to the evolving research area of environment-based digital designing. Regardless of the future changes in global and local windiness, the wind indeed significantly influences architectural designs. The interaction works both ways; the urban configuration, the rotation of the buildings concerning the wind flow, and the global and local building shape, all impact the wind conditions. These factors incorporated into the early conceptual design phase can be, among other aspects, vital players in creating sustainable architecture with excellent environmental fitness.

**In Chapter 1: Background**, a multiple-angle insight of this interdisciplinary topic is performed.

**In Chapter 2: Wind-driven design**, an application study in Sweden proves that parametrically-designed wind-driven architecture can enhance the local wind microclimate while addressing the extreme wind situations and proposing solutions to revert them. Incorporating the analysis of the specific wind situation into the form-finding of architectural shapes influences reciprocal wind-architecture interactions. The CFD analysis implemented in the early design phase leads to skilled predictions of the future design-wind microclimate interactions and the influence of the wind on the structural integrity of buildings.

**In Chapter 3: Wind-adaptive building envelope**, the designed tensegrity-membrane envelope, which dynamically responds to the instant changes in the wind velocity, direction, and changing wind loads, creates an entirely new reading of architecture by its occupants. Moreover, it enhances the wind microclimate and contributes to the reduction of wind surface suction on buildings. This structural system is a vision for shape-shifting building envelopes that can elastically respond to the wind load in real-time.

 
L. Kabošová et al., *Designing with the Wind*, Digital Innovations in Architecture, Engineering and Construction, https://doi.org/10.1007/978-3-031-24441-4_4

## 4.1 Contribution to Practice and Scientific Field

### *4.1.1 Interdisciplinary Wind-Driven Design Method*

- The wind-driven architectural design approach merges environmental engineering, architecture, and structural engineering. It enables architects to estimate the influence of their designs on the wind, already in the phase of design concepts. The improved environmental performance of architecture and urbanism created in this manner gains importance in the era of climate change.
- Combining digital parametric design with wind-performative design within one working environment and incorporating open-source CFD software into the process enables the utilization of the wind-based approach in day-to-day practice.
- The wind analysis (of existing win conditions) before the actual architectural proposal, followed by the wind-driven design, leads to utilizing the wind as a desirable phenomenon throughout the lifespan of the designed architecture.
- Drag coefficient $c_d$, wind surface pressure $p$, and the wind flow lines were obtained for complex architectural shapes. These values are crucial for designing structures of complex-shaped buildings.

### *4.1.2 Wind-Adaptive Envelope*

- Without computer control or sensors, the tensegrity-membrane wind-adaptive envelope morphs in real-time in the wind fluxes. Because the wind speed and direction fluctuate, this quick adaptation is a benefit. Real-time response is made possible by the embedded material properties and geometry-wind relations.
- The wind-adaptive envelope reacts to varying wind velocities and directions by redistributing the load, ultimately leading to lighter building structures. Moreover, due to its effects on the wind flow, it can enhance the wind comfort in the building vicinity.
- The dimples resulting from the acting wind force can significantly decrease (almost 29% for an aerodynamic shape) the wind suction on the building surface. Through the adaptive envelope, the wind itself reduces the wind suction on buildings.

## 4.2 Challenges and Future Work

### *4.2.1 Interdisciplinary Wind-Driven Design Method*

- Incorporating genetic algorithms and ML-based wind analysis into the wind-driven urbanism proposals can swiftly determine the general design direction of the project.
- The VR-based wind analysis could be exploited so the architects would see the flow influenced by their designs in the 3D space.

### *4.2.2 Wind-Adaptive Envelope*

- The most fitting flexible and, at the same time protecting textile for the membrane within the tensegrity-membrane building envelope could be designed and subsequently custom-made, specifically for adaptive envelopes.
- The cross-section of the wind-adaptive structure's components could be optimized for their efficient utilization under loads.
- Experimentations with different tensegrity-membrane shape configurations could lead to variant shape changes and thus contrasting effects on the wind flow.
- Other stimuli, including sunlight or ambient temperature, could be researched, as many factors can induce the adaptive behavior of buildings.

The manufacturer's authorised representative in the EU is Springer Nature Customer Service Centre GmbH, Europaplatz 3, 69115 Heidelberg, Germany. If you have any concerns regarding our products, please contact ProductSafety@springernature.com

Printed and bound by CPI Group (UK) Ltd, Croydon, CR0 4YY

28/09/2026

02224078-0008